T0316564

Sustainability, Innovation and Procurement

Advances in Mathematics and Engineering

Series Editor
Mangey Ram
Department of Mathematics, Graphic Era University, Dehradun, Uttarakhand, India

The main aim of this Focus book series is to publish the original articles that bring up the latest development and research in mathematics and its applications. The books in this series are in short form, ranging between 20,000 and 50,000 words or 100 to 125 printed pages, and encompass a comprehensive range of mathematics and engineering areas. It will include, but won't be limited to, mathematical engineering sciences, engineering and technology, physical sciences, numerical and computational sciences, space sciences, and meteorology. The books within the series line will provide professionals, researchers, educators, and advanced students in the field with an invaluable reference into the latest research and developments.

Recent Advancements in Software Reliability Assurance
Edited by Adarsh Anand and Mangey Ram

Sustainability, Innovation and Procurement
Edited by Sachin Kumar Mangla and Sunil Luthra

For more information about this series, please visit: https://www.crcpress.com/Advances-in-Mathematics-and-Engineering/book-series/AME

Sustainability, Innovation and Procurement

Edited by
Sachin Kumar Mangla and Sunil Luthra

CRC Press is an imprint of the
Taylor & Francis Group, an **informa** business

CRC Press
Taylor & Francis Group
6000 Broken Sound Parkway NW, Suite 300
Boca Raton, FL 33487-2742

First issued in paperback 2021

ISBN-13: 978-1-138-36548-3 (hbk)
ISBN-13: 978-1-03-217657-4 (pbk)
DOI: 10.1201/9780429430695

Publisher's Note

The publisher has gone to great lengths to ensure the quality of this reprint but points out that some imperfections in the original copies may be apparent.

Library of Congress Cataloging-in-Publication Data

Names: Mangla, Sachin K., editor. | Luthra, Sunil, editor.
Title: Sustainability, innovation and procurement / edited by Sachin Kumar Mangla and Sunil Luthra. Description: Boca Raton, FL : CRC Press/Taylor & Francis, 2020. |
Series: Advances in mathematics and engineering |
Includes bibliographical references and index.
Identifiers: LCCN 2019039831 (print) | LCCN 2019039832 (ebook) | ISBN 9781138365483 (hardback ; acid-free paper) | ISBN 9780429430695 (ebook)
Subjects: LCSH: Industrial procurement. | Sustainable development. Classification: LCC HD39.5 .S88 2020 (print) | LCC HD39.5 (ebook) | DDC 658.7--dc23
LC record available at https://lccn.loc.gov/2019039831 LC ebook record available at https://lccn.loc.gov/2019039832

Visit the Taylor & Francis Web site at
http://www.taylorandfrancis.com

and the CRC Press Web site at
http://www.crcpress.com

Contents

Preface

THE COMPLETE WORK OF this book is organized into four chapters. A brief description of each chapter follows.

In Chapter 1, Sengupta and Shukla analyze the ways in which companies implement sustainable procurement practices in the agricultural sector. This sector faces several challenges; hence, policymakers have been involved in adopting sustainable business practices to improve farm yield as well as strengthen the social welfare of marginalized farmers in emerging markets such as India. The authors have conducted a descriptive single case study of a leading organic product manufacturer in India to analyze the ways of implementing sustainable procurement processes at the field level. This chapter highlights the instrumental role of business model innovation—through the lens of the sharing economy—in incentivizing different stakeholders of the procurement process, with specific emphasis on marginalized farmers. Moreover, the authors have depicted the achievement of sustainable procurement regarding environmental and social sustainability in the supply chain structure through business model innovation. Further, the proposed framework highlights the role of innovation in removing the existing barriers to sustainable procurement in the bottom of the pyramid (BoP) market in India through an example from the organic farming industry.

In Chapter 2, Dev et al. discuss additive manufacturing (AM) processes, emerging technologies for customized products with many application areas, including in the aerospace, automobile, and medical fields. These techniques are now available globally to manufacture functional parts using metal, ceramics, polymers, and many other innovative materials. Though AM is itself the name of sustainability, its exponential development and implementation are relevant factors for the depletion of energy, materials, and other resources. The direct procurement of materials and tools also plays an important role in achieving the right balance in additive manufacturing processes. It may affect the tradeoff between goods availability, quality, and the buyer–supplier relationship. The world market status of AM is expected to exceed $21 billion in 2020, and people are planning to produce all critical and non-critical spare parts using AM technologies. This leads to the progressive convergence of natural, building and digital atmospheres, which are accountable for economic, societal, and environmental shaping. Therefore, attention to systematic planning for procurement and resource sustainability is required. This chapter provides insight into the potential of AM technology for sustainability and the procurement of innovative materials. The major aspects of sustainability and the effects of AM process parameters on energy consumption, material waste, and production time are investigated.

In Chapter 3, Agarwal and Mathiyazhagan discuss the growing concern of stakeholders towards the environmental and social impact of procurement networks, which has amplified the importance of the incorporation of sustainability in automobile spare parts (ASP) manufacturing companies in India. This has led several ASP manufacturing enterprises to explore the possibility of supply chain collaboration with suppliers to effectively manage raw materials and components. With each partner aiming at a common goal of achieving sustainability, collaboration becomes an effective strategy in reaching a desirable sustainability level under pressure from various sources, such as the government, customers, etc. In this context, the first step for the manufacturer is to understand the list of motivational factors (MFs) that put pressure on manufacturers to introduce sustainable collaboration with suppliers. Furthermore, we need to find the vital role of MFs in sustainable collaboration with the help of multi-criteria decision-making. The novelty of this chapter, therefore, lies in the use of the best–worst method (BWM) to identify the most critical MFs for the inclusion of sustainability as per the decision-maker's opinion. The study is validated for the case of a manufacturer of four-wheeler spare parts situated in northern India. The results demonstrate the usefulness of the BWM in reducing the complexity of the decision-making process while taking into consideration many qualitative and quantitative motivation factors.

In Chapter 4, Sharma et al. proclaim that the farmer is an important part of the dairy industry, regardless of country. India is the largest milk-producing nation in the world, with 18% involvement of the dairy industry in the economy. Most of the world's population depends on dairy industries for their livelihood, with about one-quarter involved in the production or purchase of milk and related products. It is essential for the Indian dairy industry to be able to fulfill growing customer demand with a satisfactory supply of milk. The procurement of milk and related products is very important in meeting this demand. Procurement is the act of gaining or obtaining goods and services. The process includes the preparation and processing of demand, along with final receipt and approval of payment. Procurement mainly depends on technology and innovation. The chapter focuses on innovative practices for the procurement of milk in the Indian dairy industry. The four practices and recommendations given are very helpful in increasing the productivity of the dairy industry in India.

Acknowledgments

The editors acknowledge the help of all the people involved in this project and, more specifically, the authors and reviewers who took part in the review process. Without their support, this book would not have become a reality.

We thank each one of the authors for their contributions. The editors wish to acknowledge the valuable contributions of the reviewers regarding the improvement of the quality, coherence, and content presentation of the chapters. Most of the authors also served as referees; we highly appreciate their double task.

We are grateful to all members of CRC Press, Taylor & Francis Group, for their assistance and timely motivation in producing this volume. We also thank Prof. Mangey Ram, Series Editor, for helping us throughout the project.

We hope that readers will share their experiences after reading our Focus book, which seeks to provide links between sustainability, innovation, and procurement.

Editors

Dr. Sachin Kumar Mangla is a lecturer in knowledge management and business decision-making at the University of Plymouth Business School, Plymouth, United Kingdom. He works in the fields of green supply chains, circular economy and sustainability, cross-disciplinary research in supply and operations management, knowledge management–based decision-making, industry 4.0, risk management, simulation, optimization, reverse logistics, renewable energy, and empirical research. He earned his doctorate specializing in operations and supply chain management from the Indian Institute of Technology (IIT) in Roorkee, India. He loves to write research papers and projects. He has published or presented several papers in reputed (ABS and ABDC indexed) national and international journals. He has an h-index score of 24, an i10-index score of 32, and more than 2000 Google Scholar citations. He received the 2017 Most Cited Paper Award for his paper "Risk analysis in green chain using fuzzy AHP approach: A case study." Recently, he edited the book *Sustainable Procurement in Supply Chain Operations*, published by CRC Press (Taylor & Francis Group). He is also currently editing several special issues in *Production Planning & Control: The Management of Operations*; *Resources, Conservation, and Recycling*; *Annals of Operations Research*; *Management of Environmental Quality*; *Journal of Resource Policy*; and *Journal of Enterprise Information Management*.

Dr. Sunil Luthra is an assistant professor at the State Institute of Engineering and Technology (formerly known as the Government Engineering College), Nilokheri, Haryana, India. He has been teaching for the past 15 years. He has contributed more than 100 research papers to national and international refereed journals and conferences at both national and international level. His scholarly work has been acknowledged in several international journals of repute such as *Journal of Cleaner Production*, *International Journal of Production Economics*, *Production Planning & Control*, *International Journal of Production Research*, *Renewable and Sustainable Energy Reviews*, *Resources, Conservation, and Recycling*, *Energy*, *Journal of Resource Policy*, and many more, and conferences of repute such as SOM-14, NITIE–POMS, AGBA, GLOGIFT 14, and GLOGIFT 15, among others. His research is also in the spotlight. His works have received more than 2000 citations (h-index = 24). His ResearchGate score is higher than 85% of other members. His specific areas of interest are operations management, green supply chain management, sustainable supply chain management, sustainable consumption and production, reverse logistics, renewable/sustainable energy technologies, and business sustainability, among others.

Contributors

Vernika Agarwal
Amity International Business School
Amity University
Noida, Uttar Pradesh, India

Saty Dev
Department of Mechanical Engineering
Motilal Nehru National Institute of Technology
Allahabad, Uttar Pradesh, India

Yaşanur Kayıkcı
Engineering Faculty
Türk Alman Üniversitesi
Istanbul, Turkey

K. Mathiyazhagan
Department of Mechanical Engineering
Amity School of Engineering & Technology
Amity University
Noida, Uttar Pradesh, India

Pravin P. Patil
Department of Mechanical Engineering
Graphic Era (Deemed to Be) University
Dehradun, Uttarkhand, India

Surya Prakash
Department of Mechanical Engineering
BML Munjal University
Gurugram, Haryana, India

Tuhin Sengupta
Department of Operations Management & Quantitative Techniques
Indian Institute of Management Indore
Indore, Madhya Pradesh, India

Yogesh Kumar Sharma
Department of Mechanical Engineering
Graphic Era (Deemed to Be) University
Dehradun, Uttarkhand, India

Suwarna Shukla
Department of Operations Management & Quantitative Techniques
Indian Institute of Management Indore
Indore, Madhya Pradesh, India

Rajeev Srivastava
Department of Mechanical Engineering
Motilal Nehru National Institute of Technology
Allahabad, Uttar Pradesh, India

Neema Tufchi
Department of Biotechnology
Graphic Era (Deemed to Be) University
Dehradun, Uttarkhand, India

Pushpendra Yadav
Department of Mechanical Engineering
Faculty of Engineering
Dayalbagh Educational Institute
Agra, Uttar Pradesh, India

CHAPTER 1

Business Model Innovation in Sustainable Procurement

A Case Study on Organic Farming in India

Tuhin Sengupta and Suwarna Shukla

CONTENTS

1.1 INTRODUCTION

Agriculture is one of the main contributors to employment for the world's population. While it is appreciated that the world economy is slowly moving toward the service sector, the agricultural sector still contributes more than 26% of global employment (World Bank, 2018a). With respect to global ranking, India occupies the third position in agricultural output. Geographically, India is home to 17% of the world's population, i.e., approximately 1.33 billion in 2017 (World Bank, 2018b). Despite the agricultural sector being one of the major contributors to the nation's economy, its contribution to the Indian gross domestic product (GDP) shows a declining trend. According to the Central Statistical Organization, the farm and agricultural sector made a 13.9% contribution in 2013–14 to India's GDP in comparison with 14.6% in 2010–11, which is consistent with the increase in the share of the service sector in GDP. Moreover, there has been a sharp decline in the overall workforce (11 percentage points) of the agricultural and allied sectors (FICCI, 2015). Despite the decline, more than 60% of rural households still depend on agriculture and more than 40% of the total labor force in India still works in the agricultural sector (FICCI, 2015).

In the agricultural sector, multiple issues result in the loss of yield. These problems include weed growth (An et al., 2018; Fahad et al., 2015; Mani et al., 1968), pests (Basak et al., 2017; Ghini et al., 2015; Savary et al., 1997), and excessive use of genetically modified crops (Fuganti-Pagliarini et al., 2017; Qaim and Zilberman, 2003). Furthermore, the existing process-driven efficiencies (traditional farming versus organic farming) (Kirchmann, 2018) and resource-driven inefficiencies (manual farming tools versus mechanized farming methods) (Geng et al., 2018; Rada and Fuglie, 2019; Zhang et al., 2017; Dai and Dong, 2014; Kislev and Peterson, 1981; Culpin, 1968) make farm productivity an emerging issue in developing nations. To mitigate such issues in the farming sector, organic farming has proven to be an important tool in enhancing the overall farm yield. Therefore, the adoption of organic farming practice can lead to considerable improvement in yield in the long run. Our chapter highlights the implementation of a structured sustainable procurement strategy through which organic farming can mitigate such issues in the farming sector.

India is one of the fastest emerging economies in the world (Singh, 2018; Mohanty et al., 2017), and the staggering growth is attributed to technological advancements and the revamping of business operations on short timescales. "Health and Wellness (HW)" is one important sector of the economy in India. The pressing need for a healthy life has ensured that consumers choose healthy food to remain fit. The market trend shows that the demand for healthy and nutritious products has gained momentum over the years. The increasing knowledge about the harmful effects of pesticides and fertilizers has provided an opportunity for the organic market segment. Most of the resulting products belong to the premium segment of the market, which is reflected in their prices. The lack of awareness about the gap between the premium customer segment and the "poor" customer segment poses an important challenge to the manufacturers and retailers of organic products.

However, organic farming is not flourishing owing to many inherent constraints in the Indian agricultural sector (Altenbuchner et al., 2018; Ditzler et al., 2018). First, farmers are not trained to operationalize the standard operating procedures of organic farming. Second, even after operationalizing the procedures, owing to their lack of habit in organic farming practice, the farmers often fail to receive organic certifications from global certifying agencies, the conditions for which are quite stringent. Third, these certifications are quite expensive and the majority of rural farmers belong to the Bottom of the Pyramid (BoP) segment; hence, it is extremely difficult for them to acquire the certifications at their own cost. Fourth, traditional farming practice has over the years given higher yields through the use of fertilizers and pesticides. Therefore, shifting to organic farming with a long term perspective of higher yield is not viable for farmers belonging to the BoP segment. For this reason, the farmers prefer to live with a short term focus, considering that the incentives are in favor of traditional farming practice.

In this chapter, we aspire to show the role of manufacturers in the supply chain in achieving sustainable procurement practice by promoting organic farming in India. Manufacturers are introducing an innovative business model similar to the sharing economy approach for incentivizing the suppliers (marginalized farmers) to adopt organic farming (sustainable procurement) practice. Therefore, we address the following research questions through a descriptive case study approach (Baxter and Jack, 2008; Eisenhardt and Graebner, 2007; Eisenhardt, 1989) for our study.

RQ 1a: How is procurement made sustainable through innovation in the agricultural sector (Sengupta and Shukla, 2019)?

RQ 1b: How do manufacturers incentivize marginalized farmers to adopt organic farming practice?

RQ 1c: What do manufacturers achieve by ensuring sustainable procurement practice in the agricultural sector?

The remaining chapter is structured as follows: Section 1.2 presents a summarized review of literature from three diverse disciplines—access-based business model innovation, Bottom of the

Pyramid, and social responsibility—to narrow down the research gap for our study. Section 1.3 presents the rationale and details for conducting a descriptive case study approach. Section 1.4 presents the analysis and findings of our chapter. Section 1.5 concludes the chapter by highlighting the research and practice implications.

1.2 LITERATURE REVIEW

To address the research gap, we developed a social responsibility lens for viewing business operations (Sodhi, 2015). We reviewed studies from three different yet connected domains, i.e., sustainability, BoP (Sharma and Jaiswal, 2018), and shared value approaches. Under the shared value approaches, we discussed the sharing economy (Belk, 2018) and connected it with the literature on BoP by highlighting access-based business model innovation as an effective tool in mitigating risks in emerging markets such as India (Schäfers et al., 2018). Subsequently, we attempted to summarize the connections among all these domains, as it would add value to the literature on sustainability.

1.2.1 Sustainability and Sustainable Procurement

Sustainability has become a tool for large corporations to instigate sustainable initiatives as a form of corporate social responsibility, meaning achieving social responsibility through business functions (Sodhi, 2015). The existing literature on sustainability encompasses all three pillars—social, environmental, and economical—that are responsible for sustainable development (Mota et al., 2015; Seuring and Müller, 2008). Operations management (OM) literature has consistently neglected the social aspects of sustainability while giving considerable importance to the environmental perspective (Carter and Rogers, 2008; Hutchins and Sutherland, 2008). The same is even true for sustainable procurement literature because procurement is normally addressed only in OM literature.

Sustainable procurement literature has been steadily gaining momentum over the last decade owing to the importance of suppliers in the overall chain (Sengupta and Shukla, 2019). However, there is a dearth of literature related to sustainable procurement in the food sector. Rimmington et al. (2006) developed the principles and performance indicators of sustainable procurement in the catering market of the United Kingdom's public sector. Hanson and Holt (2014) evaluated the sustainable food procurement of the zoos in Britain and Ireland. Park et al. (2012) presented a comparative analysis of sustainable food procurement initiatives in Canadian universities. On similar lines, Stahlbrand (2017) demonstrated the sustainability in the procurement process of universities in Canada and the UK through value-driven food supply chains. Goggins (2018) identified the crucial contextual factors within different organizations that are instrumental in shaping the overall sustainable procurement process in the entire supply chain. The Sustainable Development Commission (2002) documented the contribution of food procurement to sustainable development and highlighted the link between procurement initiatives and their effect on health. Similar work has been carried out extensively in Western European food companies (Lacroix et al., 2015). Filippini et al. (2018) presented the factors that play a vital role in improving the adoption of organic food procurement in public schools. Our study also focuses on the health and wellness segment by presenting the ways through which organic farming practice can achieve sustainable development by positively affecting the health of the community through access-based business model innovation. We have demonstrated that the procurement strategy of the manufacturer is a key pre-requisite in ensuring organic farming practices; and hence, the strategy needs to be implemented at the field level (Sengupta and Shukla, 2019).

1.2.2 The "Poor", the Bottom of the Pyramid, and Shared Value

The BoP approach suggests that multinational corporations seek to increase profits by involving "poor" or marginalized populations as suppliers and distributors in the entire supply chain (Sodhi, 2015).

Most of the grey literature has focused on the "poor" as consumers. For instance, Unilever sells small sachets of skin-color lightening products to poor consumers in developing countries (Karamchandani et al., 2011). Christensen et al. (2015) depicted the process involved in price discount strategies with the "poor" as consumers for essential products such as water filters to prevent waterborne diseases. Contrary to this, Karnani (2007) argued using economic reasoning that goods shouldn't be marketed to marginalized consumers. However, there is very little evidence on the role of "poor as suppliers", which can possibly lead to social welfare of the population in the BoP (Sodhi and Tang, 2014). Our study has extended the literature by involving "poor as suppliers" in the supply chain with the objective of achieving sustainable procurement practice through innovation. However, the redesigning of businesses through innovation is never easy, and requires considerable contextual knowledge to implement. The concept of "shared value" (Porter and Kramer, 2006) provides considerable contextual knowledge by suggesting that multinational corporations, through the improvement and redesign of the business process, could share the benefits accrued with the people belonging to the BoP, thereby recalling the overlap between the three pillars of sustainability. The concept of "shared value" provides considerable benefit to the marginalized stakeholder without compromising on generating value for the company (Coff, 1999). Therefore, we propose to see the concept of "shared value" through the access-based business model innovation in our study (Schäfers et al., 2018).

Access-based business (ABB) model innovation ensures access to goods and/or services without ownership; people can get access by paying a usage fee, which is normally less than the cost of ownership (Wiprächtiger et al., 2019; Poppelaars et al., 2018; Carroll and Buchholtz, 2012; Matzler et al., 2015). This ensures that the marginalized population is capable of consuming essential goods through a pay per use model (Sengupta et al., 2019; Belk, 2014; Catulli et al., 2013). ABB offers improved utility to consumers and/or stakeholders in the BoP, who typically face challenges arising from ownership risks and affordability issues (Karnani, 2007). The literature on access-based consumption is diverse and involves common sharing economy models such as car sharing, room sharing, etc. (Bardhi and Eckhardt, 2012; Satama, 2014). The literature on the sharing economy has gained momentum over the last few years. For instance, Hamari et al. (2016) analyzed the reasons behind the participation of consumers in collaborative consumption. Möhlmann (2015) analyzed the likelihood of co-opting the sharing economy model in the context of collaborative consumption. Heinrichs (2013) demonstrated the role of sharing economy models in achieving sustainability at workplace. On the contrary, Malhotra and Van Alstyne (2014) highlighted the negative effects of the sharing economy model on different stakeholders. Cusumano (2015) depicted the ways in which traditional firms should compete in a sharing economy setup. In our chapter, we will explain a variant of the ABB model where contextual constraints force the manufacturer to adopt a pay per use model and to incentivize the "poor as suppliers" to implementing organic farming practice in order to achieve sustainability in the supply chain. Our research design is presented in Figure 1.1.

1.3 RESEARCH METHODOLOGY

The research methodology is presented in this section into three parts. First, we provide an explanation of the rationale behind adopting a descriptive single case approach for our study. Second, we provide brief description of the case organization, its product portfolio, and its current standing in the domestic and international market. Third, we briefly explain the ways of collecting data and our method of analyzing it to deduce the requisite insights for our study. We present our research design in Figure 1.1.

1.3.1 Rationale for Single Case Study Method

The focus of our study is to answer "how" a manufacturer is able to implement organic farming among its suppliers (farmers) through business model innovation and "why" farmers need to

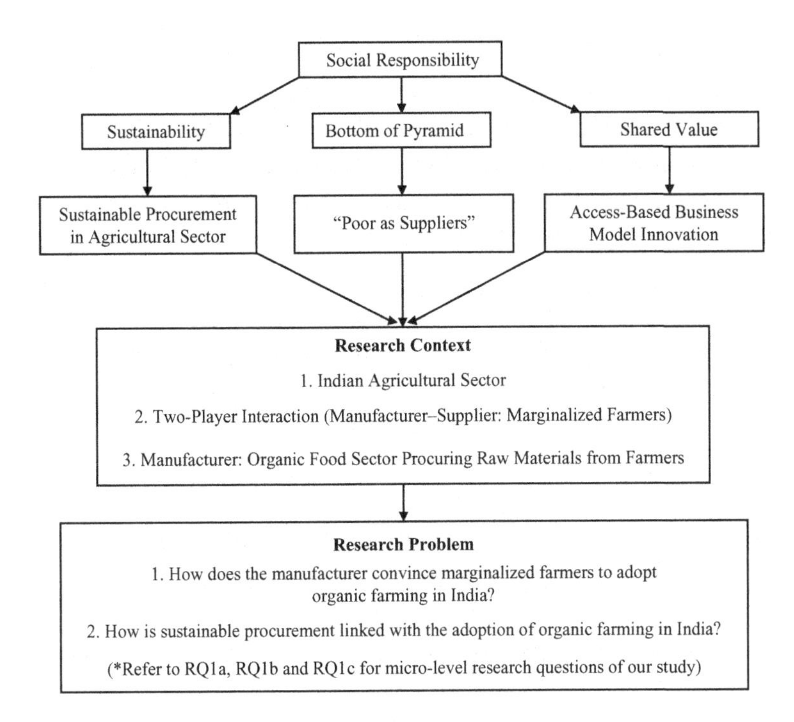

FIGURE 1.1 Research design of our study.

be incentivized to adopt such practices. To answer these questions, a case study design is the most appropriate methodology (Yin, 2003). Moreover, there are additional reasons for adopting case study design as our research methodology. First, we cannot manipulate the behavior of the actors involved in the study (Baxter and Jack, 2008). Second, we believe that it is important to highlight the contextual conditions that are relevant to the phenomenon being studied. Furthermore, it is important to decide upon the unit of analysis (Miles et al., 1994). We decided to analyze the process between manufacturer and supplier to capture the incentive mechanisms associated with both parties in the supply chain. The next step involved in the research methodology was to select the type of case study. After careful analysis and observations, it was decided that the most appropriate case research design would be a descriptive case study because, in a descriptive case study, authors typically explain and describe the intervention or phenomenon in a real life context (Baxter and Jack, 2008; Yin, 2003, Tolson et al., 2002). Depending on our understanding of the research question, we realized that it cannot be answered well through an explanatory (Joia, 2002) or exploratory (Lotzkar and Bottorff, 2001) approach within the case study. The final decision was to choose either a single case study or a multiple case study approach. Our research questions demand the interrogation of the phenomenon by analysis of the sub-units of the case. For instance, to understand the incentive mechanisms in procurement contracts, one needs to understand the contextual factors prevalent in the Indian agricultural sector and the challenges prevailing at the BoP. Therefore, it will be more prudent to conduct a within case analysis or between case analysis for our study (Ayres et al., 2003) instead of a cross case analysis (Jones et al., 2009). Therefore, we adopted a single case study approach rather than a multiple case study approach (Stake, 2013).

1.3.2 Case Organization

The organization considered in this case study is stationed in North India and is renowned for manufacturing herbal health products that are procured and produced organically at different farms. One of the major beliefs of the organization is that the whole herb is more beneficial than the extracts taken from a plant. Therefore, it is desirable that the herbs are given adequate treatment for maintaining their optimal level of potency. Consequently, the herbs grown organically are harvested manually and processed locally. To avoid photochemical breakdown, the processed herbs are dehydrated in a sterile environment and then filled into 100% vegetarian capsules devoid of any toxicity to ensure that the highest quality health products reach the customers. These products are typically sold in the USA, UK, Canada, and India. The product portfolio has three product verticals, namely, tulsi teas and infusions, herbal formulations, and packaged food.

The company is argued to be one of the first companies in the world that introduced tulsi as a product in the herbal tea category. Oxidation abundance, stress boosting, and a stronger immune system are some of the important benefits of tulsi. Various blends of tulsi tea and infusions are currently offered in the form of tea bags or in loose form in canisters, and these include sweet lemon, green tea jasmine, and honey chamomile. In addition, the certified herbal supplements possess no adverse side effects. For instance, Ashwagandha is used for stress and vitality, Breathe Free is used for respiratory relief, and Trikatu is used for digestive disorders.

1.3.3 Data Collection and Analysis Method

We conducted a systematic case study approach and identified the key stakeholders of the organization for collecting our data. Initially, we had an informal discussion with senior executives of the case organization and understood the overall business model. Subsequently, we conducted a semi-structured interaction with the key decision makers and stakeholders, which included farmers, advisors, and the marketing team. Farmers have their own land, and are currently attached to the case organization for the purpose of procurement, acting as suppliers. Farmer advisors are the typical representatives of the case organization and are given the responsibility of ensuring that the standard operating procedures are maintained at the farm level. Lastly, the marketing team is responsible for selling the final product to the market, and they also provide constant feedback to the production team. The transcripts of the semi-structured interview were analyzed and the analysis was carried out on the basis of the research question as mentioned earlier in the chapter. The research methodology framework is depicted in Figure 1.2.

1.4 ANALYSIS AND FINDINGS

We studied a scenario where the case organization is a manufacturer of organic products (organic tea in the form of tulsi tea, ginger tea, neem tea, and various different combinations of the same). The manufacturer faced serious problems in convincing

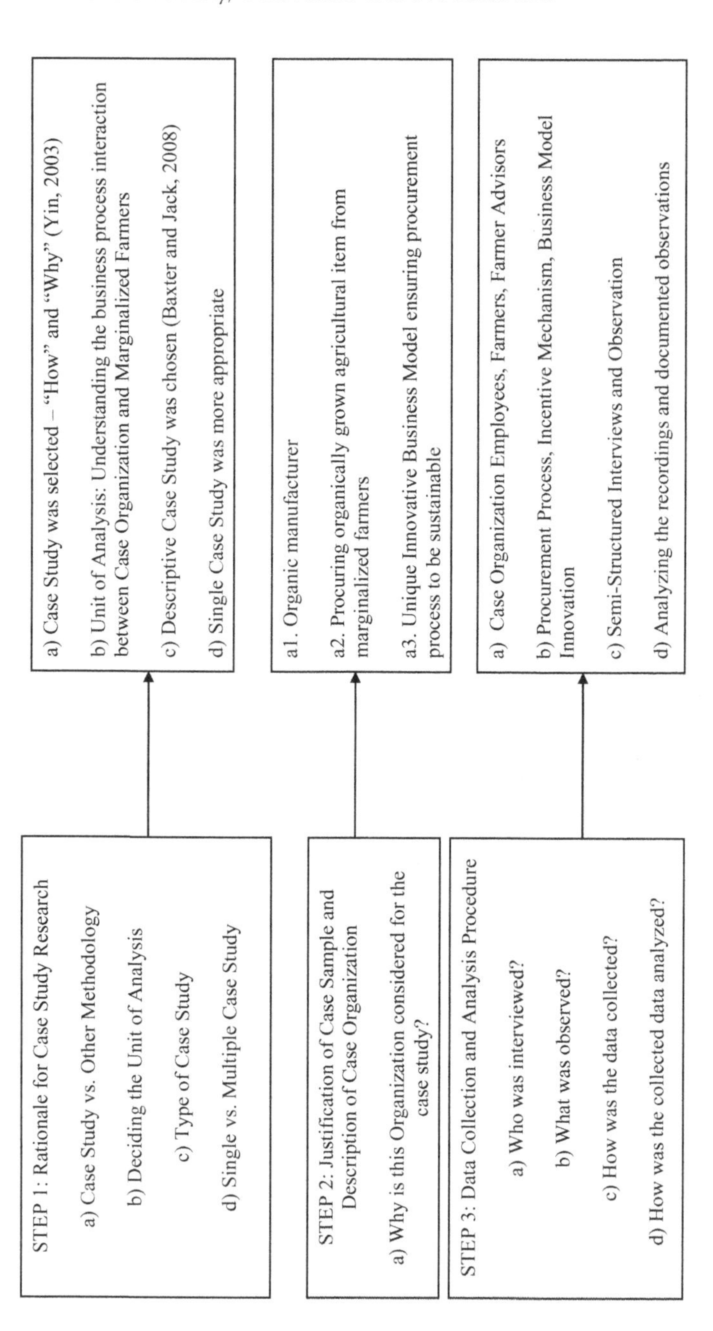

FIGURE 1.2 Research methodology framework.

farmers and farmland owners to grow agricultural products in an organic fashion. The challenge became a daunting task, as the manufacturer had to ensure the incorporation of global standard certifications on these farmlands. Considering the contextual factors in India, most farmers belong to the BoP level. Hence, farmers have shifted over the years from organic practice to traditional farming practice with the hope that farm yield would considerably increase with the use of fertilizers and pesticides, which was evident in the 1990s, following the Green Revolution in India. However, the farmers have not realized that the nutrient and fertility levels of the natural soil have been reduced because of the constant use of pesticides and fertilizers. Therefore, with the passage of each subsequent year, productivity falls at a steady pace and farmers are forced to use larger amounts of pesticides and fertilizers, which in turn further degrades the soil quality (Savary et al., 1997). Therefore, the income potential of farmers reduces in the long run because of the severe degradation of the soil nutrients. In emerging markets like India, the problem is further aggravated as farmers typically own small landholdings. The only solution to this problem is the reversal of farming practice from traditional to organic farming. This will not only improve the financial and social conditions of the farmers but will also improve the environmental quality of the soil in the long run. Further, organic foods are considered good for the end consumers in terms of health benefits. Therefore, the final challenge is to convince the farmers to adopt organic farming practice. Since farmers are the suppliers and the case organization is the manufacturer, the optimal procurement contract should be such that the farmers are incentivized to adopt organic farming practice instead of continuing to use traditional farming practice.

We found that the farmers are very reluctant to produce organic vegetables, plants in their farmland for a variety of reasons. First, there is a dearth of know-how in terms of the standard operating procedure of organic farming because they have been following traditional farming practice for the last two generations. Second, traditional farming practice has given them higher yields, and they are reluctant to believe that the use of fertilizers and pesticides will further decrease their yield by causing soil degradation. Third, they have a tried and tested model, where agri-businesses consistently buy their produce and are comfortable with their current state of affairs. Fourth, farmers are unable to gauge the market potential of organic foods and the possible manufacturers in the market because they are uncertain about the demand for this food. Fifth, it takes a lot of time to prepare the land (2–3 years) before organic vegetables and fruits can be produced in the designated area. Finally, the cost of certification and training is so high that it is beyond the reach of marginalized farmers.

Hence, it has been necessary for the case organization to come up with a different business model that could incentivize the farm owners. The case organization devised a three-part tariff system; one part is direct payment and the other two parts are the costs borne by the case organization in aiding the adoption of organic farming practice.

Let us discuss each part of the contract in detail to better understand the incentive structure. The first part involves paying a price per unit of the produce that is higher than the price paid by a competing manufacturer for goods produced through traditional farming practice. Furthermore, the case organization ensures that whatever the produce from the farm, every good is bought at the premium price even against the optimal inventory decisions of the manufacturer. This provides an initial incentive to the farmers to enter into organic farming practice. On the basis of the literature on the theory of incentives and adverse selection, it can be argued that this initiative reduces the constraint related to participation, which means that the net utility for the farmer adopting organic farming practice should be positive. Since the case organization is paying a premium price and buying all the produce, the farmers are to a certain extent willing to take the risk associated with organic farming. However, it should be noted that the case organization needs to incentivize further to keep the supply constant and increase the produce to meet future demand.

Therefore, the case organization needs to position the contract in such a way that the farmers are incentivized to provide a larger share of their farm area to the case organization. The farmlands are not owned by the case organization and so the organization shares the benefits of the produce with the farmers; hence, this model clearly depicts the sharing economy model (access-based consumption). Here, the menu of contracts is as follows: The larger the area of farmland shared with case organization, the higher the share of social and economic benefits provided to the farmers by the case organization. The social and economic benefits are the other two parts of the tariff payment discussed earlier. The first part is the certification and training costs borne by the case organization, which increases with increase in the shared farm area. It ensures that marginalized farmers are exempted from paying for global certification and training for organic farming. The second part is the social benefit in terms of medical coverage and employment to women in the farmland, which is again dependent on the shared farm area. Therefore, the case organization has employed a pay per use model (certification + training + employment + produce + premium price) for each additional hectare of land used for organic farming instead of taking any ownership of the farmland. This is exactly the variant of access-based business model innovation as mentioned in the literature, where one party accesses the product or service for a pay per use model. It can further be argued that this model is analogous to the incentive compatibility constraint in adverse selection literature, where one of the parties is incentivized to choose a type of contract from a menu of contracts. In our case, the marginalized farmers are incentivized to choose organic farming in comparison to traditional farming practice. Thus, a robust procurement policy in terms of contract design ensures the implementation of sustainable procurement practice at the field level. We present the integrated framework for successful business model innovation to promote sustainable procurement practice through the adoption of organic farming in Figure 1.3.

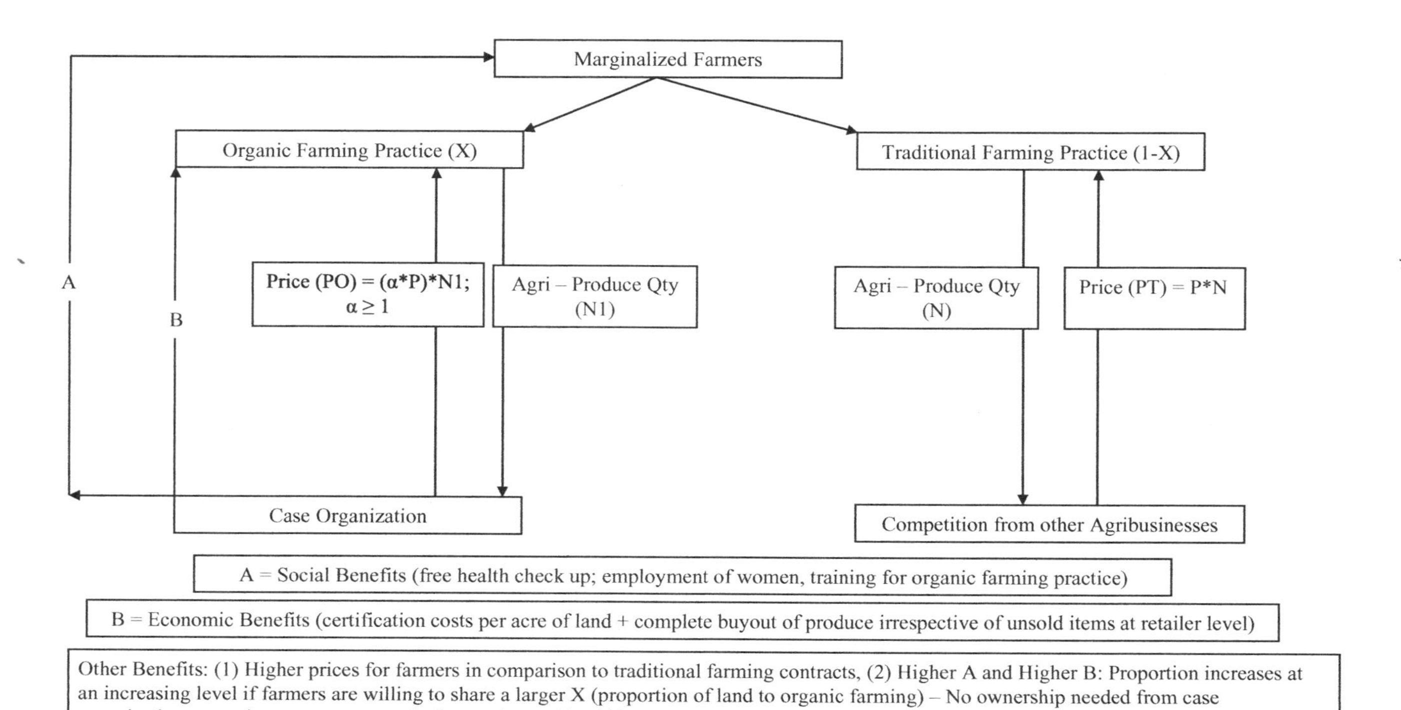

FIGURE 1.3 Integrated framework for adopting sustainable procurement in the agricultural sector.

We document the following lessons associated with the adoption of organic farming:

1. Organic farming practice has improved the livelihood of marginalized farmers, which is in accordance with premium income, health benefits, and the assurance of unsold produce in the inventory of the farmers.
2. Organic farming practice has ensured that the end consumers eat healthy food, thereby indirectly contributing to the Sustainable Development Goals.
3. Organic farming improves the quality of soil nutrients, thereby contributing toward enhancing the "green" aspect of the environment.
4. The access-based business model innovation has shown the importance of appropriate procurement contract design to ensure organic farming practice as a viable and competing option against traditional farming practice.

In a nutshell, business model innovation is instrumental in ensuring that the procurement process is sustainable, with special emphasis on the social sustainability of marginalized suppliers. This is a strong addition to the OM literature on social sustainability. Furthermore, environmental benefits accrue for soil health as well as for the end consumers, who receive healthy foods free from pesticides and fertilizers, thus marginally contributing to the environmental sustainability literature.

1.5 DISCUSSION

This chapter illustrates the use of access-based business model innovation in promoting organic farming in India with the objective of achieving sustainability in the procurement process (RQ 1a). We conducted a single case study approach and analyzed the ways in which manufacturers incentivize farmers (both economically and socially) to promote sustainable procurement practice (RQ 1b). We observed that not only farmers and the case organization but also the end consumers will benefit from the production of healthy food. We further documented the enhancement of the overall yield of the farm through the use of organic farming practice in the long run, thereby promoting ecological balance in the environment (RQ 1c). Our study directly contributes to a number of the Sustainable Development Goals (SDG 1, SDG 2, SDG 3, SDG 8, SDG 9, and SDG 12) by promoting innovation, environmental up-gradation, social benefit for the poor, and economic sustainability for multinational corporations, in addition to the supply of safe and healthy food product to their customers. This chapter provides an important framework that demonstrates the importance of social responsibility in multinational corporations (Sodhi, 2015).

1.5.1 Research Implications

Our chapter provides important research implications for diverse literature fields, encompassing innovation (Sinha et al., 2019), sustainable procurement, and industrial organization. First, this study adds to the literature on access-based business model innovation by focusing on the features of the pay per use model in relation to the agricultural sector, thereby promoting sustainable procurement practice (Schor and Fitzmaurice, 2015). Second, there are limited studies on sustainable procurement in the food sector; hence, this chapter provides a framework showing the ways of achieving social and environmental sustainability through organic farming (Filippini et al., 2018). Third, this chapter provides an analog to the literature on the theory of incentives within the umbrella of industrial organization (Laffont and Tirole, 1993). The manufacturer primarily provides a minimum incentive to the farmers for listening to the concept of organic farming. Farmers are provided with different contracts to choose from, e.g. organic farming versus traditional farming (Balakrishnan and Koza, 1993); the manufacturer has designed the incentive structure in such a manner that farmers profit from organic farming practice in comparison to traditional farming practice.

1.5.2 Practice and Social Implications

Our chapter presents key lessons for practice and policy implications. First, this study provides a strong basis for future entrepreneurs to adopt incentive-driven mechanisms for marginalized farmers in emerging markets to promote organic farming. Incentive mechanisms have been extensively studied, and these studies likely strongly validate the existing literature/theory on farming contracts (Sainathan and Groenevelt, 2019; Shen et al., 2019). Second, our study provides a framework under the umbrella of the sharing economy model for existing agricultural businesses to delve further into innovative practices for promoting sustainable procurement. Hence, this study is an addition to the scant literature on the sharing economy in the agricultural and food sector (Richards and Hamilton, 2018). Further, our study depicts the co-creation of value in favor of all stakeholders in the process. With the increase in farmers' income, the nutrient level of the soil improves, and the buyer's business model strives toward success, with the aim of providing healthy food to the customers. This is an extension to the value co-creation literature on the sharing economy, which emphasizes the need to reduce price wars and look for innovative ways to co-exist in the economy (Zhang et al., 2018). Third, our study provides key implications for agricultural policy makers in understanding the needs of the BoP population while devising a sustainable agricultural policy. In other words, agricultural policy makers have to focus on incentive-driven schemes for farmers to adopt sustainable farming practice with a long term perspective on farm yield. The adoption of sustainable farming practice may decrease the rate at which rural to urban migration is currently taking place. Moreover, the current problems of low farm yield will slowly improve in the long run, thus enabling more supply to the nation's growing population and improving the welfare of marginalized farmers.

1.5.3 Limitations

Our chapter has certain limitations, which pose opportunities for future research. First, this study does not take into account the incentives that are offered by established agriculture-based companies for the promotion of traditional farming practice. Therefore, we could not capture the impact of competition at the manufacturer level on the case organization's decision variables. Second, our study does not take into account the degree of measurable impact in terms of social welfare at the BoP level. Owing to the limitations on social sustainability literature in operations management, our study could not delve deeper into the measurable impact of sustainable procurement practice for the marginalized farmers. Third, our study could have gained better understanding through validation of the findings with the utilization of a multiple case study approach. Instead of adopting a cross case analysis to capture and compare sustainable procurement practices, multiple case organizations on similar lines could have helped to triangulate the important factors leading to the adoption of organic farming in India. However, due to the scarcity of large scale organic manufacturers in India, we restricted our study to one organization.

REFERENCES

Altenbuchner, C., Vogel, S., & Larcher, M. (2018). Social, economic and environmental impacts of organic cotton production on the livelihood of smallholder farmers in Odisha, India. *Renewable Agriculture and Food Systems, 33*(4), 373–385.

An, Y., Gao, Y., & Ma, Y. (2018). Growth performance and weed control effect in response to nitrogen supply for switchgrass after establishment in the semiarid environment. *Field Crops Research, 221*, 175–181.

Ayres, L., Kavanaugh, K., & Knafl, K. A. (2003). Within-case and across-case approaches to qualitative data analysis. *Qualitative Health Research, 13*(6), 871–883.

Balakrishnan, S., & Koza, M. P. (1993). Information asymmetry, adverse selection and joint-ventures: Theory and evidence. *Journal of Economic Behavior & Organization, 20*(1), 99–117.

Bardhi, F., & Eckhardt, G. M. (2012). Access-based consumption: The case of car sharing. *Journal of Consumer Research, 39*(4), 881–898.

Basak, S., Das, S. S., & Pal, S. (2017). Nonlinear modelling of rice leaf folder infestation on Boro rice in Pundibari (A part of Cooch Behar district). *Journal of Entomology and Zoology Studies, 5*(2), 967–972.

Baxter, P., & Jack, S. (2008). Qualitative case study methodology: Study design and implementation for novice researchers. *The Qualitative Report, 13*(4), 544–559.

Belk, R. (2014). You are what you can access: Sharing and collaborative consumption online. *Journal of Business Research, 67*(8), 1595–1600.

Belk, R. (2018). Foreword: The Sharing Economy. *The Rise of the Sharing Economy: Exploring the Challenges and Opportunities of Collaborative Consumption.* ABC-CLIO, Santa Barbara, CA, pp. ix–xii.

Carroll, A., & Buchholtz, A. (2012). *Business and Society: Ethics, Sustainability, and Stakeholder Management,* 9th edn. Cengage Learning, Stamford, CT.

Carter, C. R., & Rogers, D. S. (2008). A framework of sustainable supply chain management: Moving toward new theory. *International Journal of Physical Distribution & Logistics Management, 38*(5), 360–387.

Catulli, M., Lindley, J. K., Reed, N. B., Green, A., Hyseni, H., & Kiri, S. (2013). What is Mine is NOT Yours: Further insight on what access-based consumption says about consumers. In *Consumer Culture Theory* (pp. 185–208). Emerald Group Publishing Limited.

Coff, R. W. (1999). When competitive advantage doesn't lead to performance: The resource-based view and stakeholder bargaining power. *Organization Science, 10*(2), 119–133.

Culpin, C. (1968). *Profitable Farm Mechanization.* Harper-Collins, London.

Cusumano, M. A. (2015). How traditional firms must compete in the sharing economy. *Communications of the ACM, 58*(1), 32–34.

Dai, J., & Dong, H. (2014). Intensive cotton farming technologies in China: Achievements, challenges and countermeasures. *Field Crops Research, 155,* 99–110.

Ditzler, L., Breland, T. A., Francis, C., Chakraborty, M., Singh, D. K., Srivastava, A., … and Decock, C. (2018). Identifying viable nutrient management interventions at the farm level: The case of smallholder organic Basmati rice production in Uttarakhand, India. *Agricultural Systems, 161,* 61–71.

Eisenhardt, K. M. (1989). Building theories from case study research. *Academy of Management Review, 14*(4), 532–550.

Eisenhardt, K. M., & Graebner, M. E. (2007). Theory building from cases: Opportunities and challenges. *The Academy of Management Journal, 50*(1), 25–32.

Fahad, S., Hussain, S., Chauhan, B. S., Saud, S., Wu, C., Hassan, S., ... & Huang, J. (2015). Weed growth and crop yield loss in wheat as influenced by row spacing and weed emergence times. *Crop Protection, 71*, 101–108.

FICCI. (2015). *Transforming Agriculture through Mechanisation*. Grant Thornton India LLP, New Delhi, India.

Filippini, R., De Noni, I., Corsi, S., Spigarolo, R., & Bocchi, S. (2018). Sustainable school food procurement: What factors do affect the introduction and the increase of organic food? *Food Policy, 76*, 109–119.

Fuganti-Pagliarini, R., Ferreira, L. C., Rodrigues, F. A., Molinari, H. B., Marin, S. R., Molinari, M. D., ... & Neumaier, N. (2017). Characterization of soybean genetically modified for drought tolerance in field conditions. *Frontiers in Plant Science, 8*, 448.

Geng, Y., Yu, H., Li, Y., Tarafder, M., Tian, G., & Chappell, A. (2018). Traditional manual tillage significantly affects soil redistribution and CO_2 emission in agricultural plots on the Loess Plateau. *Soil Research, 56*(2), 171–181.

Ghini, R., Torre-Neto, A., Dentzien, A. F., Guerreiro-Filho, O., Iost, R., Patrício, F. R., ... & DaMatta, F. M. (2015). Coffee growth, pest and yield responses to free-air CO_2 enrichment. *Climatic Change, 132*(2), 307–320.

Goggins, G. (2018). Developing a sustainable food strategy for large organizations: The importance of context in shaping procurement and consumption practices. *Business Strategy and the Environment, 27*(2), 838–848.

Hamari, J., Sjöklint, M., & Ukkonen, A. (2016). The sharing economy: Why people participate in collaborative consumption. *Journal of the Association for Information Science and Technology, 67*(9), 2047–2059.

Hanson, J., & Holt, D. (2014). Sustainable food procurement in British and Irish zoos. *British Food Journal, 116*(10), 1636–1651.

Heinrichs, H. (2013). Sharing economy: A potential new pathway to sustainability. *GAIA-Ecological Perspectives for Science and Society, 22*(4), 228–231.

Hutchins, M. J., & Sutherland, J. W. (2008). An exploration of measures of social sustainability and their application to supply chain decisions. *Journal of Cleaner Production, 16*(15), 1688–1698.

Joia, L. A. (2002). Analysing a web-based e-commerce learning community: A case study in Brazil. *Internet Research, 12*(4), 305–317.

Jones, N. A., Perez, P., Measham, T. G., Kelly, G. J., d'Aquino, P., Daniell, K. A., … & Ferrand, N. (2009). Evaluating participatory modeling: Developing a framework for cross-case analysis. *Environmental Management, 44*(6), 1180.

Jones Christensen, L., Siemsen, E., & Balasubramanian, S. (2015). Consumer behavior change at the base of the pyramid: Bridging the gap between for-profit and social responsibility strategies. *Strategic Management Journal, 36*(2), 307–317.

Karamchandani, A., Kubzansky, M., & Lalwani, N. (2011). Is the bottom of the pyramid really for you. *Harvard Business Review, 89*(3), 107–111.

Karnani, A. (2007). The mirage of marketing to the bottom of the pyramid: How the private sector can help alleviate poverty. *California Management Review, 49*(4), 90–111.

Kirchmann, H. (2018). Victor M. Shorrocks: Conventional and organic farming—A comprehensive review through the lens of agricultural science. *Journal of Plant Nutrition and Soil Science, 181*(3), 471–471.

Kislev, Y., & Peterson, W. (1981). Induced innovations and farm mechanization. *American Journal of Agricultural Economics, 63*(3), 562–565.

Lacroix, R. N., Laios, L., & Moschuris, S. (2015). A model to measure SMEs sustainable procurement implementations from a study of Western European Food and Beverage Companies. *Journal of Regional & Socio-Economic Issues, 5*(3), 47–49.

Laffont, J. J., & Tirole, J. (1993). *A Theory of Incentives in Procurement and Regulation*. MIT Press, Cambridge, MA.

Lotzkar, M., & Bottorff, J. L. (2001). An observational study of the development of a nurse-patient relationship. *Clinical Nursing Research, 10*(3), 275–294.

Malhotra, A., & Van Alstyne, M. (2014). The dark side of the sharing economy… and how to lighten it. *Communications of the ACM, 57*(11), 24–27.

Mani, V. S., Gautam, K. C., & Chakraborty, T. K. (1968). Losses in crop yield in India due to weed growth. *International Journal of Pest Management C, 14*(2), 142–158.

Matzler, K., Veider, V., & Kathan, W. (2015). Adapting to the sharing economy. *MIT Sloan Management Review*, *58*(2), 71–77.

Miles, M. B., Huberman, A. M., Huberman, M. A., & Huberman, M. (1994). *Qualitative Data Analysis: An Expanded Sourcebook*. Sage.

Mohanty, S., Patra, P. K., Sahoo, S. S., & Mohanty, A. (2017). Forecasting of solar energy with application for a growing economy like India: Survey and implication. *Renewable and Sustainable Energy Reviews*, *78*, 539–553.

Möhlmann, M. (2015). Collaborative consumption: Determinants of satisfaction and the likelihood of using a sharing economy option again. *Journal of Consumer Behaviour*, *14*(3), 193–207.

Mota, B., Gomes, M. I., Carvalho, A., & Barbosa-Povoa, A. P. (2015). Towards supply chain sustainability: Economic, environmental and social design and planning. *Journal of Cleaner Production*, *105*, 14–27.

Park, B., Reynolds, L., & Initiative, P. G. (2012). Local and Sustainable Food Procurement. *Public Good Initiative*. Retrieved from http://studentfood.ca/wp-content/uploads/2012/08/Public-Good-Initiative-Local-and-Sustainable-Food-Procurement-A-comparative-analysis.pdf.

Poppelaars, F., Bakker, C., & van Engelen, J. (2018). Does access trump ownership? Exploring consumer acceptance of access-based consumption in the case of smartphones. *Sustainability*, *10*(7), 2133.

Porter, M. E., & Kramer, M. (2006). The link between competitive advantage and corporate social responsibility. *Harvard Business Review*, *84*(12), 78–92.

Qaim, M., & Zilberman, D. (2003). Yield effects of genetically modified crops in developing countries. *Science*, *299*(5608), 900–902.

Rada, N. E., & Fuglie, K. O. (2018). New perspectives on farm size and productivity. *Food Policy*, *84*, 147–152.

Richards, T. J., & Hamilton, S. F. (2018). Food waste in the sharing economy. *Food Policy*, *75*, 109–123.

Rimmington, M., Carlton Smith, J., & Hawkins, R. (2006). Corporate social responsibility and sustainable food procurement. *British Food Journal*, *108*(10), 824–837.

Sainathan, A., & Groenevelt, H. (2019). Vendor managed inventory contracts–coordinating the supply chain while looking from the vendor's perspective. *European Journal of Operational Research*, *272*(1), 249–260.

Satama, S. (2014). Consumer adoption of access-based consumption services: Case AirBnB. Master's thesis, Aalto University, Espoo, Finland.

Savary, S., Srivastava, R. K., Singh, H. M., & Elazegui, F. A. (1997). A characterisation of rice pests and quantification of yield losses in the rice-wheat system of India. *Crop Protection*, *16*(4), 387–398.

Schäfers, T., Moser, R., & Narayanamurthy, G. (2018). Access-based services for the Base of the Pyramid. *Journal of Service Research*, *21*(4), 421–437.

Schor, J. B., & Fitzmaurice, C. J. (2015). Collaborating and connecting: The emergence of the sharing economy. Reisch, L. A., & Thøgersen, J. (Eds.), *Handbook of Research on Sustainable Consumption*. Edward Elgar Publishing, Cheltenham, UK, pp. 410–426.

Sengupta, T., Narayanamurthy, G., Moser, R., & Hota, P. K. (2019). Sharing app for farm mechanization: Gold Farm's digitized access based solution for financially constrained farmers. *Computers in Industry*, *109*, 195–203.

Sengupta, T., & Shukla, S. (2019). Conceptual framework in sustainable procurement. Mangla, S. K., Luthra, S., Jakhar, S. K., Kumar, A., & Rana, N. P. (Eds.), *Sustainable Procurement in Supply Chain Operations*. CRC Press, Boca Raton, FL, pp. 3–38.

Seuring, S., & Müller, M. (2008). From a literature review to a conceptual framework for sustainable supply chain management. *Journal of Cleaner Production*, *16*(15), 1699–1710.

Sharma, G., & Jaiswal, A. K. (2018). Unsustainability of sustainability: Cognitive frames and tensions in bottom of the pyramid projects. *Journal of Business Ethics*, *148*(2), 291–307.

Shen, B., Choi, T. M., & Minner, S. (2019). A review on supply chain contracting with information considerations: Information updating and information asymmetry. *International Journal of Production Research*, *57*(15–16), 4898–4936.

Singh, R. (2018). Energy sufficiency aspirations of India and the role of renewable resources: Scenarios for future. *Renewable and Sustainable Energy Reviews*, *81*, 2783–2795.

Sinha, A., Sengupta, T., & Alvarado, R. (2019). Interplay between technological innovation and environmental quality: Formulating the SDG policies for next 11 economies. *Journal of Cleaner Production*, *242*, 118549.

Sodhi, M. S., & Tang, C. S. (2014). Supply-chain research opportunities with the poor as suppliers or distributors in developing countries. *Production and Operations Management*, *23*(9), 1483–1494.

Sodhi, M. S. (2015). Conceptualizing social responsibility in operations via stakeholder resource-based view. *Production and Operations Management*, *24*(9), 1375–1389.

Stahlbrand, L. (2017). Can values-based food chains advance local and sustainable food systems? Evidence from case studies of university procurement in Canada and the UK. *International Journal of Sociology of Agriculture & Food*, *24*(1), 77–95.

Stake, R. E. (2013). *Multiple Case Study Analysis*. Guilford Press.

Sustainable Development Commission. (2002). *Food procurement for health and sustainable development: Summary of work submitted Sustainable Procurement Group*. Sustainable Development Commission, London.

Tolson, D., Fleming, V., & Schartau, E. (2002). Coping with menstruation: Understanding the needs of women with Parkinson's disease. *Journal of Advanced Nursing*, *40*(5), 513–521.

Wipråchtiger, D., Narayanamurthy, G., Moser, R., & Sengupta, T. (2019). Access-based business model innovation in frontier markets: Case study of shared mobility in Timor-Leste. *Technological Forecasting and Social Change*, *143*, 224–238.

World Bank. (2018a). *International Labour Organization, ILOSTAT database*. Data retrieved in November 2017. https://data.worldbank.org/indicator/SL.AGR.EMPL.ZS

World Bank. (2018b). *United Nations Population Division. World Population Prospects: 2017 Revision*. https://data.worldbank.org/indicator/SP.POP.TOTL?locations=IN

Yin, R. K. (2003). Case study research design and methods third edition. *Applied Social Research Methods Series*, *5*, 1–300.

Zhang, T. C., Jahromi, M. F., & Kizildag, M. (2018). Value co-creation in a sharing economy: The end of price wars? *International Journal of Hospitality Management*, *71*, 51–58.

Zhang, X., Yang, J., & Thomas, R. (2017). Mechanization outsourcing clusters and division of labor in Chinese agriculture. *China Economic Review*, *43*, 184–195.

CHAPTER 2

Additive Manufacturing

Saty Dev, Rajeev Srivastava,
Pushpendra Yadav, and Surya Prakash

CONTENTS

2.1 ADDITIVE MANUFACTURING

Additive manufacturing (AM) provides an advanced means of rapid product development through its unique characteristics. AM is one of the fastest-developing advanced digital manufacturing techniques around the world (Wang et al., 2015). AM technology first emerged around 1987 as stereolithography, developed by 3D Systems (Petrick and Simpson, 2013). AM is also called 3D printing (3DP), solid freeform fabrication (SFF), or layered manufacturing (LM). Variability in user choice calls for customized production, and AM technology can enable this. AM technology could be strengthened by smart supply chain management, physical internet (PI) and the internet of things (IoT). A new development of AM technologies is the introduction of different digital connections that are based on IoT. An internet of things–embedded sustainable supply chain (SSC) could provide opportunities for the transition to industry 4.0 (Manavalan and Jayakrishna, 2018). AM technology is also a potential driver of supply chain management. It shortens the length and complexity of the supply chain, which improves the performance of the production system. AM positively affects the total cost from fabrication and transportation points of view. Standard products that are designed to accommodate consumers need further processing to fit the requirements of individuals (Munsch, 2017). Multi-functionality in new products has initiated requirements for many new materials such as smart materials, nanomaterial, biomaterials, functional materials, and concrete for the commercial manufacturing of actual application parts (Lee et al., 2017). AM approaches have the ability to manufacture products using metal, ceramics, polymers, and composite materials with higher dimensional accuracy, design flexibility, and optimal time from a computer-generated design file.

Though AM is itself the name of sustainability in production, random selection of parameters, material, and processing may lead to the waste of energy, materials, and time. Therefore, there is a timely need to study the sustainability aspects of additive manufacturing. The aim of this chapter is to provide an insight into the potential of AM technologies with regard to energy sustainability and ideas for the procurement of innovative materials. Few specific works have been done on the role of AM technologies for resource sustainability with innovation and procurement of suitable materials. Therefore, this study attempts to cover this literature gap. The major aspects of sustainability and the effects of AM process parameters on energy consumption, material waste, and production time are investigated. The study of effective process parameters for material and energy sustainability through AM shows originality of work. The savings in material, energy, time, and economic and societal factors are considered as sustainability aspects for additive manufacturing. The value of this study relies on the favorability of AM and its sustainability aspects.

This study scrutinizes the following research questions:

i. What is the role of AM in sustainability, innovation, and procurement of suitable materials? In detail, this chapter wants to explain the positive aspects of AM for sustainability and the link with the procurement of suitable materials.

ii. How does AM affect the triple bottom line (TBL) of sustainability (society, economy, and the environment)?

iii. What are the effects of AM technologies on supply chain management and customization?

iv. What are the major AM process parameters that influence the consumption of material, energy, and production time?

The main objectives of this study are

i. To study additive manufacturing technologies and their contribution to material and energy sustainability
ii. To provide an overview of triple bottom line sustainability, innovation, and procurement through additive manufacturing technology
iii. To discuss the global market status of additive manufacturing and smart lightweight materials
iv. To study the important process parameters which influence material and energy waste in additive manufacturing
v. To discuss the implications of the study and future directions

To achieve the aim of study the following work was undertaken:

- Study of related literature for background and research problems
- Research objectives and questions
- Identification of effective AM process parameters for sustainable manufacturing
- Study of individual effects of process parameters on energy, material, and production time
- Conclusions, limitations, and future work

2.1.1 Global Value of Additive Manufacturing

The global value of additive manufacturing (including systems, materials, and services) is rising rapidly and is expected to reach about $21 billion in 2020 (Wohlers Report, 2015). Jiang et al. (2017)

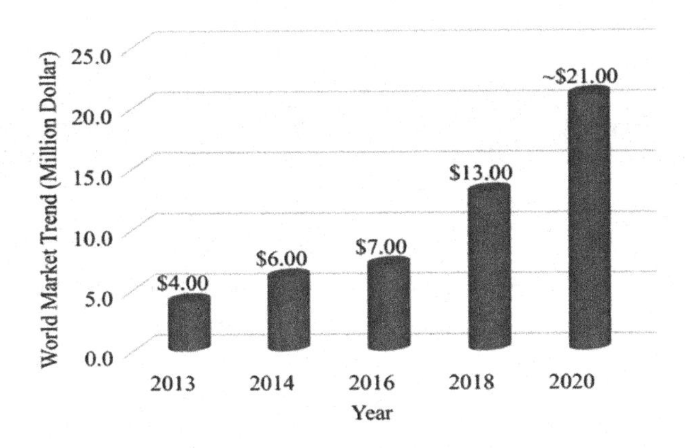

FIGURE 2.1 Global market for additive manufacturing.

estimated that after a decade, all critical and non-critical spare products will be produced through AM technologies. The large adoption of printing machines will raise the market to over $63.0 billion by 2030, as it is currently rising by approximately 30% per year. Figure 2.1 shows the strong change in the global sales of additive manufacturing. In the case of the adoption of AM, the focus of study has been on the investigation of the automobile, medical, household items, toys manufacturing, and other industries. The extreme increase in demand for AM requires a focus on the sustainability of resources.

2.1.2 Additive Manufacturing Process

Additive manufacturing is an intelligent technology where most of the production steps are taken in digital form, such as modeling with computer-aided design (CAD) software, file conversion, slicing, orientation, etc. The common steps for AM systems are shown in Figure 2.2.

2.1.3 Classification of Additive Manufacturing Technologies

According to the ASTM standard, AM techniques fall into seven main categories as shown in Figure 2.3. In any AM process, CAD software is first used to model the desired parts. The features of

FIGURE 2.2 General procedure of additive manufacturing.

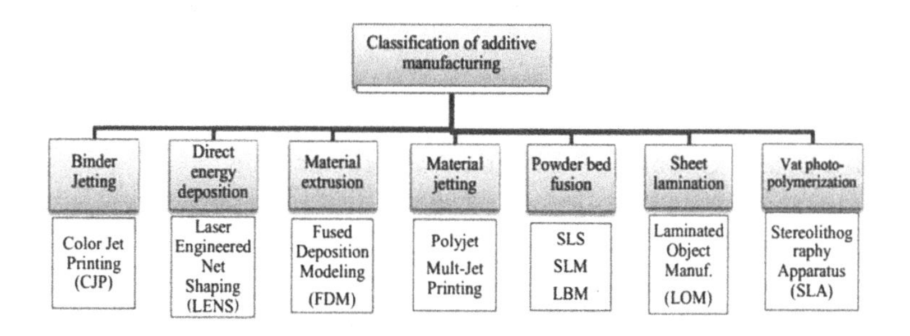

FIGURE 2.3 Classification of additive manufacturing/3D printing.

the 3D CAD files are then transferred to a file, typically with a .stl extension, which enables interfacing between the CAD model and the AM systems.

AM processes have their own merits and demerits in the fabrication of three-dimensional parts. A brief introduction to each category of AM technologies is given in the following sections.

2.1.3.1 Binder Jetting

In binder jetting (BJ), the joining of powder particles is performed with the help of a bonding agent that is in liquid form. This process of gluing powder particles together forms a three dimensional part. The bonding agent is dropped through the designed head over the powder and at the same time the platform is lowered after each layer. A typical binder jetting system has three axes, *x*, *y*, and *z*, where *x* and *y* are responsible for the horizontal position, and *z* refers to the depth. The companies ExOne and Voxeljet provide binder jetting systems for commercial purposes

(Bose et al., 2017). There is a need to understand many process parameters in binder jetting for a successful build, such as powder selection (size, shape), binder selection, printing, and post-processing. The binder jetting process does not require support and it has great design freedom with a large build size in minimum time at a low cost. This process is suitable for metal, polymer, ceramic, glass, and other materials.

2.1.3.2 Direct Energy Deposition

As the name suggests, a direct source of energy is used in direct energy deposition (DED). This energy performs multiple functions simultaneously, including deposition, melting, and solidifying the material, which is either in the form of a powder, a filament, or a wire. An automatic X-Y-Z system controls the part geometry and builds the pattern (Bose et al., 2017). Generally, the DED process utilizes a high power laser to melt the raw material. The working procedure is different from that of powder bed fusion. In the DED process, the powder is melted before being layered onto a substrate. The examples of direct energy deposition technologies include laser engineered net shaping (LENS) and 3D metal printers.

2.1.3.3 Material Extrusion

Material extrusion (ME) is one of the most widely used manufacturing techniques, in which the material is pressed out through a nozzle with constant pressure intensity. The extracted material is deposited and solidifies on the substrate. The deposition of this extracted material is carried out in such a way that the previous layer must bind with it to create a three-dimensional object. The ME process is most widely adopted for polymer materials such as acrylonitrile butadiene styrene (ABS), polycarbonate (PC), polyphenylsulfone (PPSF/PPSU), polylactic acid (PLA), thermoplastic polyurethane, etc.

2.1.3.4 Material Jetting

Material jetting (MJ) processes manufacture the three-dimensional product by depositing liquid droplets over the working platform and softening the existing layer. This layer of droplets then solidifies as a single piece casting. After the deposition and solidification of all programmed layers, the product is ejected from the working platform. An example of MJ is multijet printing machines. This process is more suitable for photopolymers.

2.1.3.5 Powder Bed Fusion

Powder bed fusion (PBF) uses thermal energy for the fusion of powder particles and a roller to smooth the powder layer. Sintering and melting are the main processes in powder bed fusion. Sintering is a typical case of biased melting. In a solid-state sintering process, the powder particles that are fused over the surface may incorporate porosity into a product, while on the other hand, fully melting and fusing the powder particles avoids porosity. Powder bed fusion is solely based on heating by a laser or electron beam. The quality of parts manufactured by PBF depends on powder flowability, which comes from appropriate powder material (Spierings et al., 2016). Selective laser sintering (SLS), selective laser melting (SLM), and electron beam melting (EBM) are examples of PBF.

2.1.3.6 Sheet Lamination

In sheet lamination (SL), material sheets are combined one over another using a CO_2 laser to create three-dimensional objects. On the basis of the bonding mechanism, SL may be categorized as adhesive joining, thermal bonding, and clamping. The strength of the manufactured part is inversely proportional to the sheet thickness (Butt et al., 2016). MCor's 3D printers are examples of this process. This process is most widely adopted for paper, plastic sheets, and metals sheets.

2.1.3.7 Vat Photopolymerization

Vat photopolymerization (VP) is generally termed a stereolithography process. In this process, photocurable resins are solidified through interaction with a laser. The reaction that takes place during solidification is known as photopolymerization, and the process includes a number of chemical compounds that introduce the required properties in the resins, for example additives, photo-initiators, and reactive monomers.

The merits and demerits of AM technologies belonging to the above seven categories are described in Table 2.1. Less waste generation, optimal geometries, lightweight components, reduced material consumption, less energy consumption, inventory waste reduction, ability to produce parts on demand, and reduced transportation burden are the major advantages of AM technologies.

Some of the available varieties of AM techniques like stereolithography apparatus (SLA), fused deposition modeling (FDM),

TABLE 2.1 Merits and Demerits of AM Technologies

Process	Merits	Demerits
BJ	• Large material variety • Ceramic molds may be fabricated • Low imaging energy	• More roughness • Lower strength • Rework needed
DED	• The material deposition rate is high • High repair efficiency	• Lower part complexity • Low surface finish • Low dimensional accuracy
ME	• Low cost • High variety of materials • Versatile and easy to customize	• Long build time • Sharp external corners cannot build
MJ	• Lower waste of material • High accuracy • High variety of material	• Rework required • More waste
PBF	• No support required • No waste • High variety of materials • High accuracy	• More surface roughness of Polymer • Low build rate • Small build size
SL	• Good fabrication speed • Support structures are not required • Low internal stress	• More material waste • Removal of support is difficult
VP	• Good accuracy and resolution • Good fabrication speed • Low specific energy	• Support required • Rework required

SLS, and 3DP are more popular for the fabrication of three-dimensional objects and commercially availability.

SLS is one of the most widely used AM processes and can produce objects with complex geometry and low volume production while consuming significantly less time than traditional processes (Xiaohui et al., 2015). SLS uses a high-temperature CO_2 laser beam to sinter the selected layers of fine powder. The laser beam strikes a specific location to fuse the powder as directed by the design. A roller is used to spread a fresh powder layer on the platform after producing each layer. The loose bed is lowered after completion of each layer by a distance equal to the thickness of that layer (10 to 100 μm) which is controlled by a piston. This process repeats until the shape of the object to be fabricated is obtained. A vacuum environment is required to protect the work from the hardships of humidity and oxidation. Commercially available SLS techniques are still constrained to producing objects that are smaller than 500 micrometers because the laser focus diameter is limited to 50–300 μm (Vaezi et al., 2013). Variables including laser power, hatch length, scan spacing, and bed temperature may influence the properties (i.e. density and hardness) of the polymer material fabricated by selective laser sintering (Singh et al., 2017). This process can be used to fabricate materials such as polymers, metals, or combinations of the two. Polymers such as acrylic styrene and polyamide (nylon) are popularly used, which have mechanical properties similar to the injected part (Wong and Hernandez, 2012).

Stereolithography apparatus was the first commercial additive manufacturing technique and was developed by 3D Systems. SLA is based on the principle of layer-by-layer polymerization of a photosensitive resin with the help of ultraviolet (UV) light (Vaezi et al., 2013). SLA is one of the most widely adopted methods due to its good quality, high accuracy, and rapid building rate. It can produce parts with a high resolution of 200 nm, along with outstanding surface finish. SLA is suitable for producing larger objects with polymers, metals, and ceramics (Bose et al., 2017). The principal components of this apparatus are a container filled with liquid photosensitive resin, a platform, and a curing UV laser.

An additive manufacturing system based on the extrusion process consists of computer-controlled deposition nozzles to develop three-dimensional objects with a specified structure and composition (Vaezi et al., 2013). FDM uses a feed of a thin filament of plastic that is melted through the print head and extruded in a specified thickness, approximately 0.25 mm. FDM is suitable for many materials such as polyphenylsulfone, polycarbonate, PC–ABS blends, ABS, etc. Sometimes the properties of parts manufactured by FDM may be improved by additional work such as vapor smoothing of an ABS product. This also improves the dimensional accuracy (Chohan and Singh, 2016). This technique possesses some unique advantages, such as that there is no need for chemical post-processing, a less costly setup, and no curing of resin, resulting in materials produced in a more cost-effective way. The FDM technique also has some demerits, such as it has low resolution and is a slower process than other AM techniques. Therefore, a separate finishing process is sometimes needed to obtain a smooth surface (Wong and Hernandez, 2012). This technique is mainly used in direct 3D printing of polymeric materials because of the low melting temperature of polymers as compared to ceramics and metallic substances (Bose et al., 2017).

The 3DP technique uses a water-based liquid binder and starch-based powder to construct the part as instructed by a CAD model. Powder particles are sprinkled on the bed and are glued together by a binder jet. This process is adaptable for a number of polymer materials (Wong and Hernandez, 2012). The technique is more preferable for three-dimensional microfabrication. The 3DP technique has some limitations like insufficient surface quality, size constraints, and porosity (Vaezi et al., 2013).

2.2 MASS CUSTOMIZATION

Additive manufacturing can produce a product on the spot per customer requirements. It eliminates the need to hold a stock of spare parts. The design of the product is completely flexible, able to produce each part in a production run at optimal manufacturing cost. This is the unique capability of AM technology that enables mass-customization. Adaptive customization can tailor, modify, or reconfigure standard products. For example, if a patient needs a novel product, his tailored data is acquired to adjust the initial design configuration to match their data. Thereafter, customized fabrication is performed through AM technology to fulfill the demands of the customer. This kind of serviceability reduces waste of the material, time, and energy used for processing. The savings of material, time and energy improve the sustainability of production. The preservation of natural resources and the environment is a very popular topic in the last decade. Sustainability is arising as a challenge that designers and engineers must work to establish in the future. In recent times, sustainability is taught as an integral part of many technical courses.

Diegel et al. estimated that, out of the world's population of nearly 7.0 billion, approximately 1.7 billion are consumers who have lifestyles devoted to the accumulation of non-essential goods. These products include bigger houses, more cars, and more consumer goods. It is also shown that more than 50% of these worldwide consumers live in developing countries (Diegel et al., 2010). There are nearly 240 million consumers in China and 120 million in India. Product design that incorporates sustainability improves the financial imperatives, integrities, and other aspects of overall sustainability. This strategy aims to use ecological principles to establish the triple bottom line of sustainability. AM technology potentially influences the supply chain from a fabrication viewpoint. An agile supply chain could be adopted easily with the use of AM technology, and this reduces the cost for manufacturing firms by up to 75% (Rathee et al., 2017). Additive manufacturing allows the direct fabrication of parts for customized needs (Petrick and Simpson, 2013).

2.3 AM TECHNOLOGIES AND SUPPLY CHAIN MANAGEMENT

Additive manufacturing technology regarded as a technological driver in supply chain management. AM's adoption in engineer-to-order processes affects the configuration of supply chains in many ways. The addition of new members such as AM machine vendors or customers, and the elimination of some existing members such as material suppliers, show the effect on the supply chain (Oettmeier, 2016; Özceylan et al., 2018). Narain and Sarik (2006) highlighted that an RP/AM system will be able to serve customers at their site for extended supply chain functions. AM technology improves supply chain performance and offers strong value by the following paths: (1) The companies seek AM technologies to provide the value in supply chain rather than radical alterations (Marchese and Crane, 2018); (2) Companies use the opportunities enabled by AM technologies to take advantage of economies of scale; (3) The complete alteration of supply chain and products is not possible without AM technology, which the firm needs for competitive advantage. When AM technology is combined with the internet of things (IoT), it converts a business to a smart business. The following points show the strength of AM technologies when combined with IoT:

- The best enclosures for IoT items are made possible through AM technology.
- IoT helps collect data to create a product customized by additive manufacturing.
- The factories are able to make true the shape of industry 4.0 with the help of AM processes.

AM technology adoption reduces the complexity and length of the supply chain network. Marchese and Crane (2018) depicted the effect of AM adoption on supply chain structure. They found that AM technology improves the performance of the traditional supply chain.

2.4 ADDITIVE MANUFACTURING AND SUSTAINABILITY (AMS)

The report of the United Nations Brundtland Commission defines sustainability as "development which meets the needs of the present without compromising the ability of future generations to meet their own needs" (Giret et al., 2015). The additive manufacturing saves resources of material, energy, and time, which are the main aspects of sustainable production. The triple bottom line, shown in Figure 2.4, categorizes the sustainability aspects into three classes: environmental, economic, and social sustainability. If a production process produces a product with maximum values of these three aspects (i.e. it is eco-friendly, profitable, and favorable for society) is termed as having ideal and sustainable production.

An understanding of TBL explains the current position of your business in the economy and its potential to exist into the future. It is a framework that evaluates the performance of your product/process in three dimensions: environmental, economic, and social. The TBL dimensions are also called "the three Ps": People,

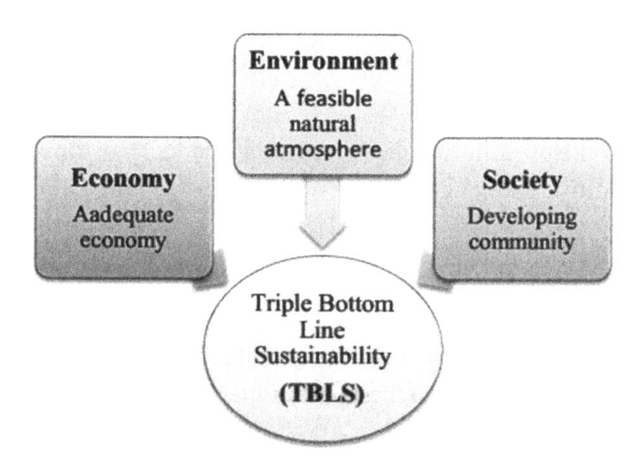

FIGURE 2.4 Aspects of the triple bottom line of sustainability.

1984	•Emergance of additive manufacturing
1992	•Early study on sustainable production
1995	•Eearly study on environmental impacts of AM
2000	•Sustainable manufacturng first appear
2010	•Attention on sustainability issues in AM
2011	•Sustainable additive manufacturing of food items
2012	•First prosthetic jaw is printed and implanted
2013	•3D printing" in USA's Union speech
2020	•Research focus on intelligent material for AM
2025	•Additive manufactuirng of cars and full human body organs
2030	•Seroous isues of energy and suitable maerials due to producion of all critical & non critical parts through AM

FIGURE 2.5 A timeline of sustainability in additive manufacturing.

planet, and profit. The historical development of AM technology and sustainability is shown in Figure 2.5.

Though sustainable development has been studied previously, greater attention toward sustainability in AM processes has been witnessed in the last decade. Within the space of two decades, sustainability in manufacturing has been achieved.

TBL sustainability in an AM process can be achieved by managing the drivers of a sustainable environment, economy, and society. Low resource consumption, minimum waste, and reduction of pollutants are the potential drivers of a sustainable environment. The economy requires improved productivity, reduced costs, and market evaluation for sustainable AM processes. Society needs benefits in businesses, qualitative production, public agreement, and ethics. Figure 2.6 explains the majors of triple bottom line sustainability of additive manufacturing.

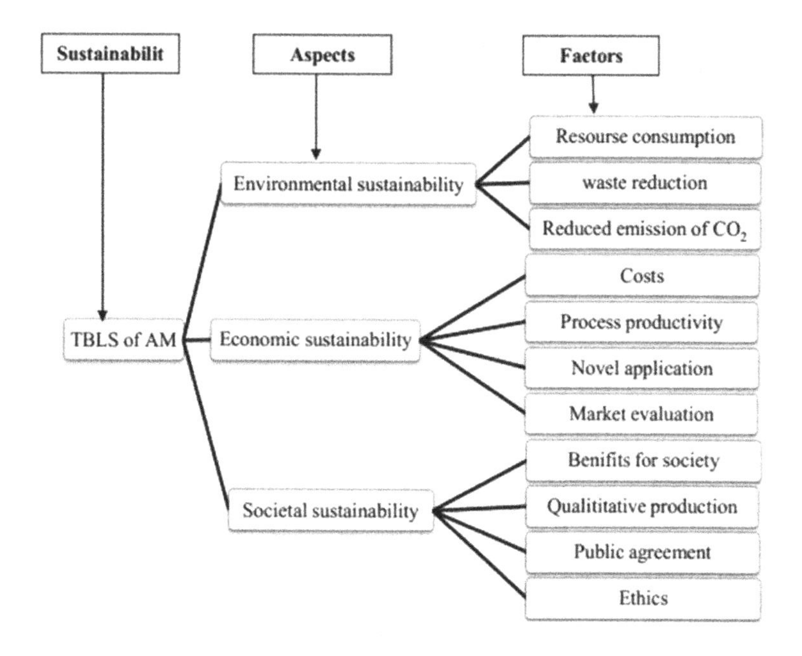

FIGURE 2.6 Measures of TBL sustainability in additive manufacturing.

2.4.1 Environmental Sustainability

AM technology is also a contributor to a pollution-free environment. Environmental sustainability means managing, reducing, and monitoring consumption, wastes, and emissions. The environmental factors considered while designing manufacturing methods should include aspects of the sensible use of natural resources. These resources include fresh air and water, energy, plant life, organic material, and land available for use. Chavarría-Barrientos et al. also wrote that AM technology is an environmental sustainability–enabling technology (Chavarría-Barrientos et al., 2018). There are many factors which measure the impact of a project on environmental sustainability, including

- Resource consumption
- CO_2 emission during manufacturing
- Amount of nitrogen oxides and sulfur dioxide produced

- Selected priority pollutants
- Waste reduction
- Compact size

2.4.2 Economic Sustainability

Based on the principle of TBL, economic sustainability is not a traditional corporate concern. It is the measure of the impact of strategy on the economic environment. Additive manufacturing technology strengthens the economy by reduced production cost of customized parts, material consumption, and process time. The economic variables deal with the bottom line and the flow of money. Economic sustainability is governed by incomes and expenditures, taxes, business drivers, employment, and business diversity factors. The examples of economic sustainability factors are

- Individual income
- Level of churn rate
- Growth of jobs
- Cost of customized parts
- The lifecycle of a product
- Extra features of product complexity

2.4.3 Societal Sustainability

The social aspect of the sustainability of AM methods is the measure of their impact on society while they are producing new products, that is, they should do so without causing any loss to society. It can increase by producing products that have good features and are attractive and favorable for human health. The social bottom line questions the belief that the less a business pays its workforce, the longer it can afford to operate. The societal measures of additive manufacturing are related to social aspects such as health, quality of life, and social capital. The examples of societal sustainability factors are

- The demand for additively manufactured parts
- Labor required
- Customization of product at the user's home
- Production time
- Health-related issues
- Capital required

AM is the potential tool of tomorrow that must be understood today to assure sustainable manufacturing (Narain and Sarik, 2006). AM technologies require raw material to create the part and in some cases a small amount of material for support, hence they are highly material-efficient processes. AM shortens the supply chain and shifts work from physical goods to digital designs. This shift enhances supply chain dynamics by reducing the "time-to-market" and hence the transport burdens of the supply chain are expected to fall (Sreenivasan et al., 2010). The AM manages the buy-to-fly ratios which relate raw material requirements to the amounts of material in final products. This ratio commonly occurs as 20:1. AM enables a buy-to-fly ratio of almost 1:1 and thus induces a noteworthy decrease in resource burden and manufacturing-related waste (Gebler et al., 2014). Considering carbon emissions, the following advantages environmental and sustainability arise with the adoption of additive manufacturing:

- Reduced amount of raw material required which leads to minimizing the material preparation time.
- Reduces energy-intensive and wasteful processes.
- Minimizes the energy consumed to design and manufacture complex geometries.

- High strength to weight ratio reduces carbon emissions by transport vehicles.
- Customized parts could be manufactured closer to the point of consumption.

All of the above points show the potential of AM to save resources and the environment, which helps to improve the sustainability of manufacturing.

2.5 ADDITIVE MANUFACTURING MATERIALS AND PROCUREMENT

Today, a variety of AM materials are available to use in the form of metals, alloys, ceramics, bioactive glass, polymers, and their combinations. A general classification of AM materials is indicated in Figure 2.7.

- Digital and smart materials include shape memory polymers (SMP), which may be processed by VP with functional applications in actuators, sensors, jewelry, and grippers.
- Ceramic materials such as UV curable monomers are used for thermal protection and are also suited by VP.
- Electronic materials like silver nanoparticles in antenna emitters and thin-film transistors, quantum dot material

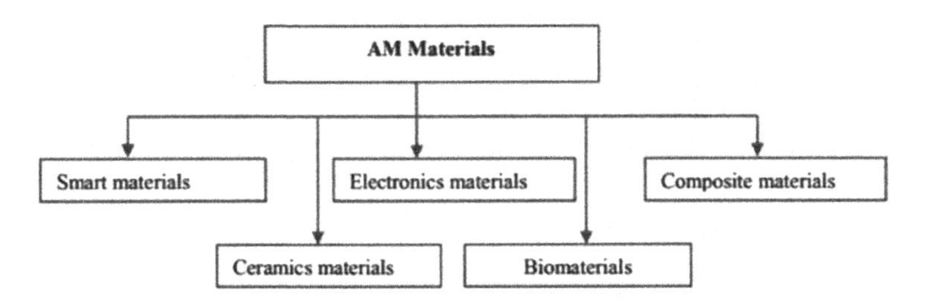

FIGURE 2.7 General classification of AM materials.

for light-emitting diodes (LEDs), and some conductive polymers for resistors are processed by material jetting.

- Biomaterials such as hydrogels suitable for tissue engineering and functional inks used in cardiac micro-physiological devices can be processed by ME.
- Composite materials such as CB/PCL, VeroWhitePlus, Tango Black Plus, barium titanate, nanoparticle/polyethene, and glycol diacrylate are used in sensors, fracture-resistant composites, and in the fabrication of 3D piezoelectric polymers, and can be manufactured using material extrusion, material jetting, and vat photopolymerization.

A variety of materials are also developing day by day for additive manufacturing of functional parts. The world market for advanced lightweight materials is predicted to reach around $196,299 million in 2022 whereas it was estimated $144,613 million in 2015. These lightweight materials are useful for a variety of applications including microelectronic circuits, fuel cells, sensors, and coatings against corrosion. Most of the applications of the 3D printing market are fulfilled by polymer materials (Wohler's Report, 2018). Figure 2.8 shows the market trend of advanced lightweight materials for additive manufacturing.

The major issues in additive manufacturing are suitable material development for innovative applications and the capability of

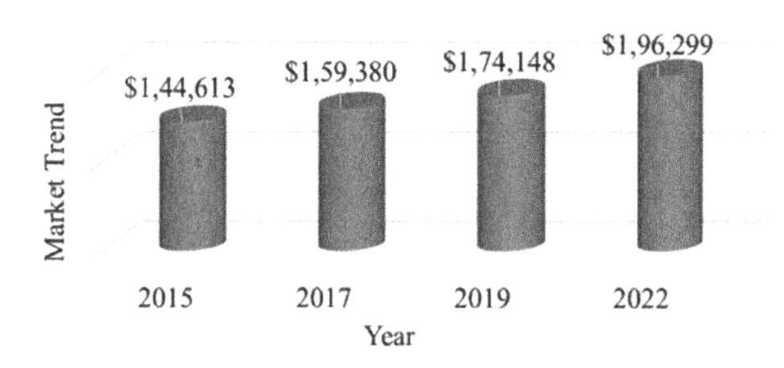

FIGURE 2.8 Global market for intelligent materials.

AM systems to print these materials. There is a temperature limit for all AM systems and the suitability of a particular material varies. Extrusion-based AM technology is generally used with materials that have a melting temperature of around 200°C.

The process of procurement involves different necessary activities for obtaining better service or products at optimal cost. The product or service that can provide the required quality is the main concern in procurement. These products or services may include raw material, technical equipment, officer equipment, testing and training, transportation, and many more. Procurement is the core component of firm's corporate strategy. The procurement of materials and different advances in AM technologies are the keys to its development worldwide. From the procurement standpoint, AM technology has major cost benefits over traditional manufacturing technologies. However, some limitations also exist, such as the availability of suitable intelligent materials for advanced applications, or the availability of AM systems that can print high entropy alloy (HEAs) materials. The size capability of AM machines is also a limitation. Additive manufacturing has the ability to meet customer requirements at the right time without the need for extra inventory.

Additive manufacturing technology enables the fabrication of smart and innovative materials for advanced applications. AM has an ability for multi-directional production and free form fabrication. Therefore, AM could be an effective process for advanced and composite materials (Yakout and Elbestawi, 2017). Functionally graded materials (FGM) and many other lightweight advanced composite materials are produced through AM processes. FGMs are advanced material used for applications where varying properties are needed. The method for manufacturing FGMs with varying composition is additive manufacturing technology (Muller et al., 2013); (Bobbio et al., 2017). AM technology is helpful in the strategic procurement of new and advanced materials for different applications such as in the aerospace industry. The main components of strategic procurement are strategy, process performance, skilled people, and innovation. AM is an advanced digital technology that can manage all requirements of the procurement of innovative materials.

2.6 PROCESS VARIABLES OF ADDITIVE MANUFACTURING

The most familiar additive manufacturing technology is FDM. In this section, the effects of extrusion-based AM process parameters on material sustainability, energy sustainability, and product quality are discussed. There are four classes of process parameters which are found to affect performance (Rathee et al., 2017):

- Operation-specific parameters
- Modeler-specific parameters
- Geometry-specific parameters
- Material-specific parameters

Operation-specific parameters concern operating characteristics such as slice height, raster width, head movement, envelope temperature, contour width, contour filling, and air gap. These variables are in the control of user. Modeler-specific parameters are related to the shape, size, and dimensions of different hardware. Examples of modeler-specific variables include nozzle diameter, filament size and feed rate, roller speed, etc. Geometry-specific parameters relate to the body of the product and the supports used to print it. Geometry-specific parameters include support structures, orientation, etc. Material-specific parameters include physical characteristics, binders, viscosity, chemical composition, and flexibility.

2.6.1 AM Process Parameters and Material Sustainability

Though additive manufacturing is itself the name of sustainability, minor issues in some AM technologies (such as extrusion-based processes) waste material and energy for extra supports.

Support volume is a general concern in all additive manufacturing processes because it leads to material and energy waste. Part build orientation is one of the most significant factors which affects the volume of support material required. Material waste can be minimized by the selection of optimal orientation (Ezair et al., 2015; Taufik and Jain, 2013; Zhang et al., 2017; Liu et al., 2018).

Verma and Rai (2016) conducted a study for part level and slice level optimization for material and energy saving. Optimal build orientation saves the material based on the part level analysis. Srivastava et al. experimentally found various parameters including orientation, contour width, raster angle, raster width, and slice height are significant for minimizing support volume, model material volume, and build time (Srivastava et al., 2017). A recent practical study conducted by Normale et al. aimed to minimize support volume and its effect on surface quality in fused deposition modeling. They proposed a new strategy to optimize the support by modification and integration in CAD software. The results found were compared with the findings of other software in terms of printing time and material required for support generation. The findings of the study show that extra support is responsible for material waste and it can be minimized through a good set of parameters (Normale et al., 2018).

With the intention of saving material, Jin et al. (2017) optimized processes and process parameters considering geometric and surface features of the final object. However, this condition compromises the mechanical strength of the product by reducing internal dimensions due to the reduction in material volume. The most influential and important process parameters for reducing material consumption in AM processes are build orientation, air gap, infill pattern, contour width, raster angle, raster width, slice height, and tool path. An optimal build orientation can save up to 40.5% of the material required for generating supports in the AM process.

2.6.2 AM Process Variables and Energy Sustainability

One of the most focused-on topics in today's manufacturing area is energy consumption (Tang et al., 2016). The saving of energy and time in additive manufacturing is always a concern. It is clear from many studies that additive manufacturing processes offer many advantages over conventional manufacturing. However, waste in the form of unnecessary processes still exists (Rejeski et al., 2018). The energy per part in AM is higher than conventional processes (Huang et al., 2013; Yoon et al., 2014). Morrow et al. (2007) compared the electrical energy consumption for direct metal deposition and machining processes. They also observed the effect of solid-to-cavity volume ratio on electrical energy consumption. A higher ratio indicates minimum energy consumption in machining while a low ratio shows lower energy consumption and emissions in direct metal deposition. The energy consumption for a PBF process during the manufacturing stage was evaluated by Sreenivasan and Bourell (2010). The study concentrated on environmental impact.

There is a large scope for reducing energy consumption through additive manufacturing. Most studies concentrate on electrical energy consumption during the fabrication stage only. It is also desirable to calculate the electrical energy used during other stages such as pre-processing and post-processing. Simultaneous multi-part production through by saves energy and manufacturing time. The optimal selection of process parameters also contributes to saving energy and other inputs through the minimization of non-value added operations.

2.6.3 AM Process Parameters and Production Time

Time is one of the major factors in manufacturing (Faludi et al., 2015). Unnecessary movement for operation and idle state of system consume extra energy and time. By removing the idle state of the system and unnecessary movements, time and energy can be saved (Faludi et al., 2016). Han et al. commented that the build time (BT) directly influences productivity (Han et al., 2002). Another study by Han et al. (2003) attempted to resolve the issue of productivity through the reduction of build time. Jin et al. stated the build time as the most important factor in many scientific and industrial applications (Jin et al., 2014). Some industries enhanced the production process to save electrical energy by selecting the best suitable processes for raw material and machining operations (Munoz and Sheng, 1995). Life cycle analysis (LCA) is a way to analyze the waste of energy and time throughout the cycle of a product (Hallstedt et al., 2018).

Time management is itself an energy- and cost-saving process. Reduced pre-processing and post-processing are themselves time-saving activities. The optimization of tool path and idle position of the machine can also minimize the build time in additive manufacturing technologies. Discontinuities and uncertainty of operational process add the unproductive time. Therefore, minimization of discontinuities may also be a solution for time and energy saving.

2.6.4 AM Process Variables and Product Properties

Mechanical properties are very important for any physical component. A number of experimental studies have been conducted to analyze the effects of various parameters on the mechanical properties of AM products when the concentration is towards material and energy saving. Ahn et al. investigated the effect of build orientation on parts produced by FDM using ABS material. The selected parameters are model build orientation, temperature, air gap, material color, and road width (diameter of bead extruded through the nozzle). Build orientation and air gap were found to have a significant effect on mechanical strength whereas other parameters have a minor effect (Ahn et al., 2002). The authors estimated the tensile strength, impact strength, and flexural strength by considering part orientation, layer thickness, air gap, and raster width as the control parameters. Impact strength increases as the only layer thickness increase out of selected parameters. Croccolo et al. conducted an experiment to analyze the mechanical properties of ABS-M30 produced by FDM. The authors investigated the effect of specimen radius, the number of contours, and longitudinal rasters. The effect of the radius on the fracture strength of ABS-M30 at different radii was evaluated (Croccolo et al., 2013).

Many studies focused on the effect of process variables on product quality. Amit Khatri produced samples of different parameters (infill percentage, raster orientation, number of shells, and axis orientation) using the MakerBot Replicator 2X. The prepared samples were further subjected to tensile testing with Shimadzu's Universal Testing Machine (UTM). The results indicated that the fracture strength of the ABS component was affected by many parameters. The fracture strength at 0° was found to be the highest and it was inversely proportional to the raster angle (Amit Khatri, 2017). Durgun and Ertan conducted an experimental study with the aim of investigating FDM technology for improving the mechanical properties of a product. Part orientation and raster angle were selected as input process parameters to investigate the mechanical properties and support volume. The results showed reduced manufacturing time and costs (Ertan, 2014). The mechanical properties of the part were affected by raster layer in the direction along the cross-sectional area of the specimen (Reley et al., 2011; Magar et al., 2018). This summary indicates that AM techniques have the potential to support distributed manufacturing and customized production while minimizing overall cost, energy, and time consumption. Mechanical properties vary according to build style, raster angle, and build orientation. The most influential parameters for the mechanical properties of an AM product are layer thickness, air gap, raster angle, road width, build pattern, and build orientation. Minimal layer height provides better mechanical strength of product but increases the manufacturing time. In material extrusion processes, the mechanical properties are poor in the direction perpendicular to the build layer.

2.7 CONCLUSIONS

All of the benefits of additive manufacturing are not easy to forecast accurately for sustainability. It requires calculations of return on investment (ROI) and sustainability aspects, which poses challenges. AM is the potential tool of tomorrow that must be understood today in order to ensure sustainable manufacturing (Narain and Sarik, 2006). As discussed in earlier sections, AM technology is the driver of sustainability and innovation for complex and lightweight geometries. Customization forces the procurement of innovative materials and intricate designs. The implementation of AM technology reduces material waste and energy consumption, and allows complexity in geometries. These technologies fulfill the custom requirements of users with a minimal cost, which is one of the drivers of societal sustainability. Now we can also propose that the study contributes toward helping managers and engineers take decisions for production prototypes and small manufacturing runs of high-value products in medical field such as dental implants, healthcare aids and devices, etc. Possible manufacturing processes can be identified for service parts, high-value form parts, products related to fashion that have a high volume/short lifespan profile, mass-produced fast-moving consumer goods with variety, etc.

The following points were observed in the course of this study.

- AM technology seems to possess immense potential for sustainability and innovation in many application areas including the aerospace, automobile, medical, education, and military fields.
- AM technology avoids the expenses required for inventory management. It provides the product at the right time with optimal quality.
- The different AM technologies have their own merits and demerits for providing quality and the sustainable production of goods. If the technology can provide the required strength to the product, it can create energy intensive products/parts and save resources, further supporting sustainable production.
- Natural resources could be saved through the optimal setting of process parameters and selection of suitable materials and machines for a particular application. It saves energy, material, and time in the manufacturing of products.

- The ability of AM technology to produce any shape motivates the development of intelligent materials and designs for many application areas. It is, of course, the need that generates the invention.
- AM technologies have the potential to support the triple bottom line of sustainability through the saving of material, energy, and time and hence cost, quality, and demand.
- However, there are many issues existing with AM technologies regarding mechanical properties of the product, post-processing, availability of intelligent materials, and the capability of AM systems.

Every study has boundaries, so too does this work. The observed potential of AM for sustainability has not been analyzed. Many of the issues found in this study can be analyzed and resolved in real-world practices for innovative applications.

REFERENCES

Ahn, S., Montero, M., Odell, D., Roundy, S. and Wright, P.K. (2002). "Anisotropic material properties of fused deposition modeling ABS", *Rapid Prototyping Journal*, 8 (4), 248–257.

Amit Khatri, A.A. (2017). "Effect of raster orientation on fracture toughness properties of 3d printed abs materials and structures", in *Proceedings of the International Mechanical Engineering Congress and Exposition (IMECE2016)*, November 11–17, 2016, Phoenix, AZ, USA, pp. 1–7.

Bobbio, L.D., Otis, R.A., Borgonia, J.P., Dillon, R.P., Shapiro, A.A., Liu, Z.K. and Beese, A.M. (2017). "Additive manufacturing of a functionally graded material from Ti-6Al-4V to Invar: Experimental characterization and thermodynamic calculations", *Acta Materialia*, 127, 133–142.

Bose, S., Ke, D., Sahasrabudhe, H. and Bandyopadhyay, A. (2017). "Additive manufacturing of biomaterials", *Progress in Materials Science*, 93, 45–111.

Butt, J., Mebrahtu, H. and Shirvani, H. (2016). "Strength analysis of aluminium foil parts made by composite metal foil manufacturing", *Progress in Additive Manufacturing*, 1 (1–2), 93–103.

Chavarría-Barrientos, D., Batres, R., Wright, P.K. and Molina, A. (2018). "A methodology to create a sensing, smart and sustainable manufacturing enterprise", *International Journal of Production Research*, 56 (1–2), 584–603.

Chohan, J.S. and Singh, R. (2016). "Enhancing dimensional accuracy of FDM based biomedical implant replicas by statistically controlled vapor smoothing process", *Progress in Additive Manufacturing*, 1 (1–2), 105–113.

Croccolo, D., De Agostinis, M. and Olmi, G. (2013). "Experimental characterization and analytical modelling of the mechanical behaviour of fused deposition processed parts made of ABS-M30", *Computational Materials Science*, 79, 506–518.

Diegel, O., Singamneni, S., Reay, S. and Withell, A. (2010). "Tools for sustainable product design : Additive manufacturing", *Journal of Sustainable Development*, 3 (3), 68–75.

Ertan, I.D.R. (2014). "Experimental investigation of FDM process for improvement of mechanical properties and production cost", *Rapid Prototyping Journal*, 20 (3), 228–235.

Ezair, B., Massarwi, F. and Elber, G. (2015). "Orientation analysis of 3D objects toward minimal support volume in 3D-printing", *Computers and Graphics*, 51, 117–124.

Faludi, J., Baumers, M., Maskery, I. and Hague, R. (2016). "Environmental impacts of selective laser melting: Do printer, powder, or power dominate?," *Journal of Industrial Ecology*, 21, S144–S156.

Faludi, J., Bayley, C., Bhogal, S. and Iribarne, M. (2015). "Comparing environmental impacts of additive manufacturing vs. traditional machining via life-cycle assessment", *Rapid Prototyping Journal*, 21 (1), 1–52.

Gebler, M., Uiterkamp, A.J.M.S. and Visser, C. (2014). "A global sustainability perspective on 3D printing technologies", *Energy Policy*, 74, 158–167.

Giret, A., Trentesaux, D. and Prabhu, V. (2015). "Sustainability in manufacturing operations scheduling: A state of the art review", *Journal of Manufacturing Systems*, 37, 126–140.

Hallstedt, S., Schulte, J. and Watz, M. (2018). "Additive manufacturing from a strategic sustainability perspective", *Proceedings of the 15th International Design Conference - Design 2018*, Dubrovnik, Croatia, 1381–1392.

Han, W., Jafari, M.A., Danforth, S.C. and Safari, A. (2002). "Tool path-based deposition planning in fused deposition processes", *Journal of Manufacturing Science and Engineering*, 124 (2), 462.

Han, W., Jafari, M.A. and Seyed, K. (2003). "Process speeding up via deposition planning in fused deposition-based layered manufacturing processes", *Rapid Prototyping Journal*, 9 (4), 212–218.

Huang, S.H., Liu, P. and Mokasdar, A. (2013). "Additive manufacturing and its societal impact : A literature review", *The International Journal of Advanced Manufacturing Technology*, 67, 1191–1203.

Jiang, R., Kleer, R. and Piller, F.T. (2017). "Predicting the future of additive manufacturing : A Delphi study on economic and societal implications of 3D printing for 2030", *Technological Forecasting and Social Change*, 117, 84–97.

Jin, Y., Du, J. and He, Y. (2017). "Optimization of process planning for reducing material consumption in additive manufacturing", *Journal of Manufacturing Systems, The Society of Manufacturing Engineers*, 44, 65–78.

Jin, Y.A., He, Y., Fu, J.Z., Gan, W.F. and Lin, Z.W. (2014). "Optimization of tool-path generation for material extrusion-based additive manufacturing technology", *Additive Manufacturing*, 1, 32–47.

Lee, J., An, J. and Chua, C.K. (2017). "Fundamentals and applications of 3D printing for novel materials", *Applied Materials Today*, 7, 120–133.

Liu, J., Gaynor, A.T., Chen, S., Kang, Z., Suresh, K., Takezawa, A., Li, L., et al. (2018). "Current and future trends in topology optimization for additive manufacturing", *Structural and Multidisciplinary Optimization*, 57, 2457–2483.

Magar, S., Khedkar, N.K. and Kumar, S. (2018). "Review of the effect of built orientation on mechanical properties of metal-plastic composite parts fabricated by additive manufacturing technique", *Materials Today: Proceedings*, 5 (2), 3926–3935.

Manavalan, E. and Jayakrishna, K. (2018). "A review of Internet of Things (IoT) embedded sustainable supply chain for industry 4.0 requirements", *Computers & Industrial Engineering*, 127, 925–953.

Marchese, K. and Crane, J. (2018). "3D opportunity for the supply chain; Additive manufacturing delivers", available at: https://www2.deloitte.com/insights/us/en/focus/3d-opportunity/additive-manufacturing-3d-printing-supply-chain-transformation.html.

Morrow, W.R., Qi, H., Kim, I., Mazumder, J. and Skerlos, S.J. (2007). "Environmental aspects of laser-based and conventional tool and die manufacturing", *Journal of Cleaner Production*, 15, 932–943.

Muller, P., Mognol, P. and Hascoet, J.Y. (2013). "Modeling and control of a direct laser powder deposition process for Functionally Graded Materials (FGM) parts manufacturing," *Journal of Materials Processing Technology*, 213 (5), 685–692.

Munoz, A.A. and Sheng, P. (1995). "An analytical approach for determining the environmental impact of machining processes", *Journal of Materials Processing Technology*, 53 (3–4), 736–758.

Munsch, M. (2017). *15 - Laser Additive Manufacturing of Customized Prosthetics and Implants for Biomedical Applications, Laser Additive Manufacturing*, Elsevier Ltd., available at:https://doi.org/10.1016/B978-0-08-100433-3.00015-4.

Narain, R. and Sarik, J. (2006). "Strategic justification of rapid prototyping systems", in E.A. Nasr and A.K. Kamrani, eds., *Rapid Prototyping: Theory and Practices*, Springer, Boston, MA, 253–270.

Normale, E., Rennes, S. De, Bretagne-loire, U. and Richir, S. (2018). "Support optimization for additive manufacturing: application to FDM", *Rapid Prototyping Journal*, 24 (1), 69–79.

Oettmeier, K. (2016). "Impact of additive manufacturing technology adoption on supply chain network structures – An exploratory case study", *Journal of Manufacturing Technology Management*, 27 (7), 944–968.

Özceylan, E., Çetinkaya, C., Demirel, N. and Sabırlıo, O. (2018). "Impacts of additive manufacturing on supply chain flow: A simulation approach in healthcare industry", *Logistics*, 2, 1–20.

Petrick, I.J. and Simpson, T.W. (2013). "3D printing disrupts manufacturing how economies of one create new rules of competition", *Research-Technology Management*, 56 (6), 12–16.

Rathee, S., Srivastava, M., Maheshwari, S. and Noor, A. (2017). "Effect of varying spatial orientations on build time requirements for FDM process : A case study", *Defence Technology*, 13 (2), 92–100.

Rejeski, D., Zhao, F. and Huang, Y. (2018). "Research needs and recommendations on environmental implications of additive manufacturing", *Additive Manufacturing*, 19, 21–28.

Riley, W., Sturges, L. and Morris, D. (2011). *Mechanics of Materials, I6th Ed.*, John Wiley & Sons, Hoboken, NJ.

Singh, S., Sachdeva, A. and Sharma, V.S. (2017). "Optimization of selective laser sintering process parameters to achieve the maximum density and hardness in polyamide parts", *Progress in Additive Manufacturing*, 2 (1–2), 19–30.

Spierings, A.B., Voegtlin, M., Bauer, T. and Wegener, K. (2016). "Powder flowability characterisation methodology for powder-bed-based metal additive manufacturing", *Progress in Additive Manufacturing*, 1 (1–2), 9–20.

Sreenivasan, R. and Bourell, D.L. (2010). "Sustainability study in selective laser sintering – An energy perspective", *Minerals, Metals and Materials Society/AIME*, Warrendale, PA, pp. 257–265.

Sreenivasan, R., Goel, A. and Bourell, D.L. (2010). "Sustainability issues in laser-based additive manufacturing", *Physics Procedia*, 5 (1), 81–90.

Srivastava, M., Maheshwari, S., Kundra, T.K. and Rathee, S. (2017). "Multi-Response Optimization of Fused Deposition Modelling Process Parameters of ABS Using Response Surface Methodology (RSM) – Based Desirability Analysis", in *5th International Conference of Materials Processing and Characterization (ICMPC 2016) Multi-Response*, 4, pp. 1972–1977.

Tang, Y., Mak, K. and Zhao, Y.F. (2016). "A framework to reduce product environmental impact through design optimization for additive manufacturing", *Journal of Cleaner Production*, 137, 1560–1572.

Taufik, M. and Jain, P.K. (2013). "Role of build orientation in layered manufacturing : A review", *International Journal of Manufacturing Technology and Management*, 27 (1/2/3), 47–73.

Vaezi, M., Seitz, H. and Yang, S. (2013). "A review on 3D micro-additive manufacturing technologies", *International Journal of Advanced Manufacturing Technology*, 67 (5–8), 1721–1754.

Verma, A. and Rai, R. (2016). "Sustainability-induced dual-level optimization of additive manufacturing process", *International Journal of Advanced Manufacturing Technology*, 88 (5–8), 1945–1959.

Wang, X., Gong, X. and Chou, K. (2015). "Review on powder-bed laser additive manufacturing of Inconel 718 parts", *Journal of Engineering Manufacture*, 231 (11), 1–14.

Wohlers Report. (2015). "*3D Printing and Additive Manufacturing State of the Industry Annual Worldwide Progress Report*", Wohlers Associates, 2015.

Wohlers Report. (2018). "*Advanced Materials and Additive Manufacturing State of the Industry Annual Worldwide Progress Report*", Wohlers Associates, 2018.

Wong, K.V. and Hernandez, A. (2012). "A review of additive manufacturing", *ISRN Mechanical Engineering*, 2012, 1–10.

Xiaohui, S., Wei, L., Pinghui, S., Qingyong, S. and Qingsong, W. (2015). "Selective laser sintering of aliphatic-polycarbonate/hydroxyapatite composite scaffolds for medical applications", *The International Journal of Advanced Manufacturing Technology*, 81, 15–25.

Yakout, M. and Elbestawi, M. (2017). "Additive manufacturing of composite materials: An overview", in *Proceedings of the 6th International Conference on Virtual Machining Process Technology*, Montréal, Canada.

Yoon, H., Lee, J., Kim, H., Kim, M., Kim, E. and Shin, Y. (2014). "A comparison of energy consumption in bulk forming, subtractive, and additive processes : Review and case study", *International Journal of Precision Engineering and Manufacturing-Green Technology*, 1 (3), 261–279.

Zhang, Y., Bernard, A., Harik, R. and Karunakaran, K.P. (2017). "Build orientation optimization for multi-part production in additive manufacturing", *Journal of Intelligent Manufacturing*, 28 (6), 1393–1407.

CHAPTER 3

Evaluation of Motivational Factors for Sustainable Supplier Collaboration in Indian Industries

Vernika Agarwal and K. Mathiyazhagan

CONTENTS

3.1 INTRODUCTION

The rising issues of climate change, resource dependency, and sustainable development are drawing more and more attention from various international bodies such as the United Nations Environment Programme, independent organizations such as the Global Reporting Initiative (GRI), and various non-government organizations. Sustainability has thus become one of the most critical concepts for research. Since the inception of sustainable development (SD) in the United Nations Conference on the Human Environment in Stockholm in 1972, there has been tremendous pressure on manufacturing firms across the globe to consider ecological and societal factors in their business (Mathiyazhagan et al. 2019). In the Indian context, the manufacturing industries are specifically in the limelight due to the rise in pollution levels and are being pressured to incorporate sustainability into their operations. In this view, suppliers play the most important role in achieving sustainability (Govindan et al. 2017). To attain sustainability in operations, the first step is to generate successful business alliances and have common sustainability goals with other channel partners, especially with suppliers (Kara and Firat 2016), as suppliers can affect the overall creation of sustainable supply chain (Amindoust et al. 2012).

Hence, a number of automobile spare parts (ASP) manufacturing enterprises are exploring the possibility of supply chain collaboration with the suppliers to more effectively manage raw materials and components. The first step in this direction is to understand the list of motivational factors (MFs) that put pressure on integrating sustainable collaboration with suppliers. The literature on collaborative alliance has mainly focused on enhancing environmental sustainability by focusing on relations between firms and non-governmental organizations (NGOs) and between firms and the government in so-called "public–private partnerships" (Niesten et al. 2017). The literature is still lacking in understanding the MFs, which is the novelty of the present chapter.

The present study addresses the MFs in the implementation of a collaborative alliance with suppliers. The detailed aims of the present study are as follows:

- To identify the MFs for the implementation of collaborative alliance with suppliers
- To identify the importance of each MF by evaluating factors using best–worst multi-criteria decision-making analysis

3.2 LITERATURE REVIEW

Collaboration is an alliance between various firms to work together and access complementary resources to share risks and generate more rewards (Simatupang and Sridharan 2002; Sandberg 2007). Traditionally, companies collaborating to reduce their costs and create a competitive advantage (Cao and Zhang 2011). In addition, resource sharing, which involves mutually sharing the responsibility of both tangible and intangible elements, is considered an important criterion for collaboration between firms (Ramanathan and Gunasekaran 2014). Collaborative alliance is also a powerful tool for enhancing the overall performance of the supply chain (Van der Vaart and van Donk 2008; Nyaga et al. 2010, Kache and Seuring 2014). Different researchers have identified different reasons for the integration of collaborative alliance with various channel members of the supply chain network. It can be demonstrated that there are various MFs that form the basis of a cooperative alliance based on the requirements of the firms.

The growth of sustainability in measuring its competitive position influences the collaborative behavior of the firm. There is little focus on research into sustainable collaboration with suppliers. The focus of studies has mostly been on ecological and economic collaboration. Klassen and Vachon (2003) in their study emphasized that the major factor for collaborative alliance between a firm and its supplier was to reduce environmental impact. This was further verified by Vachon and Klassen (2006), who examined the propensity for collaborative alliance and found that it works better when the suppliers and firm work toward a common goal of long term environmentally sound operations. Large and Thomsen (2011) analyzed the potential enablers for the performance of green supply chains and established that the degree of green supplier assessment and the level of green collaboration exert direct influence on environmental performance. Overall it can be seen in the literature that the environmental perspective is one of the MF for a collaborative alliance between supplier and firm. Another essential MF for collaborative alliance is social sustainability. A study by Hsueh (2015) analyzed the effect of corporate social responsibility (CSR) collaboration on firm and supplier profits. Moreover, research by Sancha et al. (2016) established that when the firm has higher expectations from its stakeholders for the inclusion of sustainability, it automatically enforces the same on the supplier. Despite the literature on collaboration partnerships in the supply chain, very few have discussed the MF behind the decision-making framework. Furthermore, the majority of the literature is concerned with identifying one or more factors rather than prioritizing the MF for a successful supplier–firm alliance. The concept of sustainability has not much been touched upon in the available literature; the focus is only on a single aspect of sustainability rather than all three. The present paper aims to analyze MFs that influence sustainable collaborative alliance between the firm and suppliers.

3.3 PROBLEM DESCRIPTION

The present chapter addresses issues for sustainable integration along with supply chain management in an ASP manufacturing company, which is in the National Capital Region (NCR), India. Based on government eco-friendly policies, companies are under pressure to think about sustainable production. In this view, suppliers play the most important role in achieving sustainability. Many positive factors (MFs) are pushing the companies toward sustainable collaboration. However, companies are struggling to identify the important MFs for attaining sustainable collaboration with suppliers. The company under consideration, XYZ Ltd., is one of the leading manufacturers of a broad range of wiring harnesses, battery cables, wiring sets, connectors, terminals, and interior plastic components. It provides connective solutions to each vertical of the automotive world and off-road vehicles. It has various 2/3 wheeler, commercial and utility vehicle, construction equipment, tractor, and 4-wheeler automobile manufacturing companies as its customers. The company has plants all over India. The present study focuses on the Greater Noida plant (which is around 10–12 years old). In this pursuit, the preliminary step is to understand the list of MFs that put pressure on integrate sustainable collaboration with suppliers, which is the motivation behind the present work.

Hence, the research questions addressed in the present study include

- What are the MFs that are needed for choosing suppliers for collaborative partnership?
- How should these MFs be prioritized so that the sustainability goals and objectives of the company are met?

3.4 RESEARCH METHODOLOGY

Under multi-criteria decision-making (MCDM) approaches, best–worst method (BWM) is a practical recent approach for comparing the factors by a pairwise comparison matrix. This method is better than the analytical hierarchy process (AHP). It was proposed by Jafar Rezaei in 2015. The aim of this method is to compute the priorities of factors with relatively less information, compared with other MCDM techniques (Rezaei 2015). BWM provides priority vector based on only two comparison vectors as opposed to full pairwise matrix that is utilized in AHP.

The steps of BWM are as follows (Rezaei 2015):

Step 1: Select the criteria set.

In this step, based on the literature survey and discussion with the decision-makers (DMs), the factors for comparison $F = \{f_1, f_2, ..., f_n\}$ are finalized.

Step 2: Identify the best and the worst factors.

Based on discussions with the DMs, the best and the worst factors are selected.

Step 3: Calculate the preference of the best factor over the others.

A score of 1–9 is utilized to compute the preference of the best criteria over the others based on the inputs given by the DMs. This generates the "Best-to-Others" vector as given below:

$$k_B = \left(k_{B1}, k_{B2} .., k_{Bn}\right)$$

where k_{Bi} gives the preference of the best criteria B over the ith attribute and $k_{BB} = 1$.

Step 4: Calculate the preference of the factors over the worst criteria.

This step is similar to step 3, but here, the vector that is calculated is "Worst-to-Others" as given as:

$$K_W = \left(k_{1W}, k_{2W} .., k_{nW}\right)^T$$

where k_{iW} gives the preference of the ith criteria over worst criteria W and $k_{WW} = 1$.

Step 5: Compute the optimal weights of factors.

In this step, we compute the optimal weighting vector denoted by the factors $\left(y_1^*, y_2^* \ldots, y_n^*\right)$.

The optimal weight of the ith criteria is the one which meets the following requirements: $\frac{y_B^*}{y_i^*} = k_{Bi}$ and $\frac{y_i^*}{y_W^*} = k_{iW}$

To satisfy this condition, the maximum absolute differences $\left|\frac{y_B}{y_i} - k_{Bi}\right|$ and $\left|\frac{y_i}{y_W} - k_{iW}\right|$ need to be minimized for all factors.

Thus, as given by Rezaei (2015), we can calculate the optimal weights for factors by using the following programming problem:

$$\min_i \max \left\{ \left|\frac{y_B}{y_i} - k_{Bi}\right|, \left|\frac{y_i}{y_W} - k_{iW}\right| \right\}$$

Subject to

$$\sum_n y_i = 1 \qquad \text{(P1)}$$

$$y_i \geq 0 \qquad \forall i = 1, 2, \ldots, n$$

Problem P1 is equivalent to the following linear programming formulation P2:

min η
Subject to

$$\left|y_B - k_{Bi} y_i\right| \leq \eta \qquad \forall i = 1, 2, \ldots, n$$

$$\left|y_i - k_{iW} y_W\right| \leq \eta \qquad \forall i = 1, 2, \ldots, n$$

$$\sum_i y_i = 1 \qquad \text{(P2)}$$

$$y_i \geq 0 \qquad \forall i = 1, 2, \ldots, n$$

On solving Problem P2, we get the value of η^* optimal weights $\left(y_1^*, y_2^* \ldots, y_n^*\right)$ are determined.

Step 6: Check the consistency of the solution.

Consistency of the solution is checked by calculating the consistency ratio:

$$\text{Consistency Ratio} = \frac{\eta^*}{\text{Consistency Index}}$$

The consistency index is taken from Table 3.1, given by Razaei (2015).

If the value of the consistency ratio (CR) is closer to 0, the solution is considered as more consistency while a value closer to 1 shows less consistency. When the CR is not equal to 0, it can be said that the pairwise comparison matrix is not fully consistent and we may have multiple optimality. Hence, we need to find the optimal intervals for the weights of criteria as given by the study by Rezaei (2016). The lower and upper bounds of the weights of the ith factor are determined by solving Problems P3 and P4, respectively:

$$\begin{aligned}
&\min \ y_i \\
&\textit{Subject to} \\
&\left|\frac{y_B}{y_i} - k_{Bi}\right| \le \eta^* \qquad \forall i = 1,2,\ldots,n \\
&\left|\frac{y_i}{y_W} - k_{iW}\right| \le \eta^* \qquad \forall i = 1,2,\ldots,n \qquad\qquad \text{(P3)} \\
&\sum_i y_i = 1 \\
&y_i \ge 0 \qquad\qquad \forall i = 1,2,\ldots,n
\end{aligned}$$

TABLE 3.1 Consistency Index Table for BWM

k_{Bi}	1	2	3	4	5	6	7	8	9
Consistency index (max)	0.00	0.44	1.00	1.63	2.30	3.00	3.73	4.47	5.23

$$\max\ y_i$$

$$\textit{Subject to}$$

$$\left|\frac{y_B}{y_i}-k_{Bi}\right| \le \eta^* \qquad \forall i=1,2,\ldots,n$$

$$\left|\frac{y_i}{y_W}-k_{iW}\right| \le \eta^* \qquad \forall i=1,2,\ldots,n \tag{P4}$$

$$\sum_i y_i = 1$$

$$y_i \ge 0 \qquad \forall i=1,2,\ldots,n$$

By solving these two programming problems P3 and P4, we get the optimal weights of the criteria are determined in the form of intervals. The center of these intervals can be used to rank the factors.

3.5 CASE ILLUSTRATION

3.5.1 Research Design

The purpose of the research design phase is to identify and prioritize the MFs for ASP manufacturing companies situated in Greater Noida (as discussed in detail in section 3.3). Primarily, we identify the various MFs taking into consideration the social, environmental, and economic aspects that form the basis of sustainability. The present study targeted academicians (professors from management and engineering departments) and experts from the ASP manufacturing company (managers, higher and middle level, and engineers, higher level) from the NCR to identify the MFs that are needed to reach a desirable sustainability level in their production process. An expert panel was formed, consisting of two academicians from engineering, one academician from management, and five senior managers from the ASP manufacturing company to discuss and finalize the MFs. Based on the consensus of the panel, nine MFs were identified, as given in Table 3.2.

TABLE 3.2 List of MFs for Sustainable Supplier Collaboration in Indian ASP Industries

Motivational Factors	References
Market pressure on industries towards sustainable collaboration (MF1)	Kushwaha and Sharma (2016)
Information and resource sharing to attract customers (MF2)	Cao et al. (2010), Ramanathan and Gunasekaran (2014)
Government reward and incentives schemes for achieving sustainable goals (MF3)	Morgan et al. (2016)
Continued pressure from NGOs to reduce waste (MF4)	OWN
Mutual understanding of the sustainability goals between firm and partners (MF5)	Sakka et al. (2011)
To attract worldwide customers (MF6)	OWN
Pressure from customers for eco-friendly products (MF7)	Chen et al. (2015)
Pressure from stakeholders to include human resource policies (MF8)	Zhang et al. (2017)
To provide safe and healthy working conditions (MF9)	Govindan et al. (2019)

Once we finalized the MFs, the relevant data was collected using the opinions of the expert panel. For the data collection, the DMs were asked to select the best and worst MF and give reference comparison values on a 1–9 scale. The final data collected from various DMs in the expert panel was compiled into a single piece of data by taking an average. This data was used to generate weights using the BWM technique as elaborated in section 3.4. The following section discusses in detail the application of BWM in prioritization of MFs for the inclusion of sustainability in procurement operations.

3.5.2 Application of BWM for the Case

The data collected from the expert panel was used to illustrate the methodologies and to develop a framework for identifying the vital MFs for sustainable collaboration with suppliers. As discussed above, we finalized the MFs with the help of an expert panel. A total of nine MFs were finalized for the study, based on inputs by experts from both academia and industry. Following the steps of BWM, the next step after finalization of criteria is to determine the best and worst criteria among the MFs. The best MF was found to be "MF4: Continued pressure from NGO to reduce waste" while the worst MF was "MF2: Information and resource sharing to attract customers," based on the consensus of the expert panel. Following the next step, we need to give the preference rating of best criteria over all other criteria and give the preference rating of the other criteria over worst criteria on a scale of 1–9. These pairwise comparisons are shown in Table 3.3.

The weights are determined using Problem P2, which is coded and solved using LINGO, an optimization software program. Graphically, the weights are shown in Figure 3.1.

Based on the value of η^* we calculate the CR. Here, we get the value of $\eta^* = 0.0746$, and the value of CR is given as 0.014, which is closer to 0. Thus, the solution is considered to be consistent.

TABLE 3.3 Pairwise Comparison for the MFs

Best to Others	**MF1**	**MF2**	**MF3**	**MF4**	**MF5**	**MF6**	**MF7**	**MF8**	**MF9**
MF4	2	9	3	1	6	5	4	8	3
Others to the Worst	**MF2**								
MF1	6								
MF2	1								
MF3	5								
MF4	9								
MF5	3								
MF6	4								
MF7	7								
MF8	2								
MF9	5								

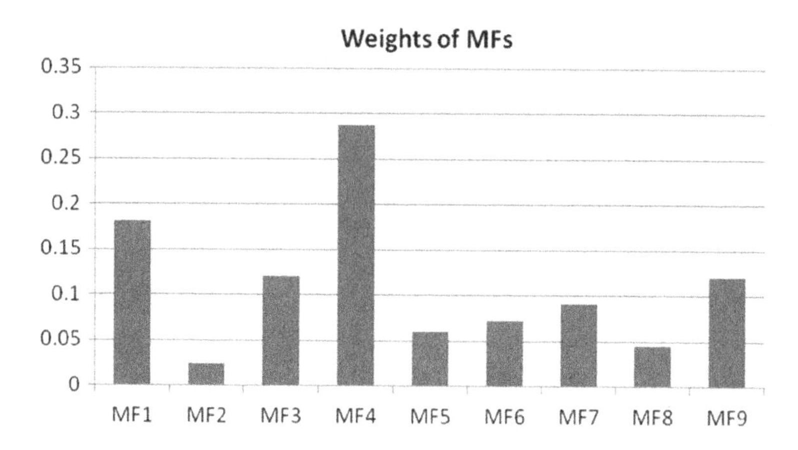

FIGURE 3.1 Weights of MFs.

3.6 RESULTS AND DISCUSSIONS

With the help of a BWM approach, the MFs were analyzed and it was found that "Continued pressure from NGOs to reduce waste" (MF4) is the foremost motivator based on the opinions of academicians and industry experts. Creating awareness of waste management is a primary concern in reducing waste and increasing sustainability (Gazzola et al. 2019). This factor is particularly important in ensuring greater pressure on the supplier to implement sustainable practices. Similarly, "Market pressure on industries towards sustainable collaboration" (MF1) is the next best factor for sustainable supplier collaboration. Due to the scarcity of sources and global warming, customers are focusing on environmentally friendly products which will increase pressure on industries to think about sustainability criteria along with traditional criteria for the supplier selection. On another side, "Information and resource sharing to attract customers" (MF2) is the worst motivational factor for sustainable supplier collaboration. From industry experts' and academicians' perspectives, information and resource sharing do not hold much importance for sustainability among suppliers out of the nine final MFs. Information sharing alone does not give more motivation compared with MF1. "Pressure from stakeholders to include human resource policies" (MF8) is the next worst MF. Similarly, "Mutual understanding of the sustainability goals between firm and partners" (MF5) and "To attract worldwide customers" (MF6) are the next on the list of worst MFs.

3.7 MANAGERIAL IMPLICATIONS

ASP manufacturing companies in India are under the strict surveillance of the government and NGOs to ensure they are including ecological and social welfare concerns in their business operations. These manufacturers are seeking the help of their suppliers by collaborating with them to effectively include sustainability aspects. The present study indicates the major MFs that are influencing this decision for ASP manufactures. There are various theoretical and policy suggestions that can be derived from the present work, which are discussed in the following sections.

3.7.1 Theoretical Relevance

Here some suggestions are made for implementing collaborative alliance with suppliers for Indian ASP manufacturers:

- Continued pressure from NGOs is the primary factor that will force the manufacturer to work in partnership with the suppliers.
- By collaborating with suppliers, ASP manufacturers can attain more international visibility in terms of recognition and awards.
- The manufacturers can make strategic training programs jointly with the help of suppliers to enhance the overall sustainability performance and to promote healthier, more sustainable supply chain networks.

3.7.2 Policy Suggestions

In lieu of the growing government pressure, there is a dire need for the manufacturers to collaborate with the suppliers to enhance the overall sustainability of their value chains. To expand upon this issue, some policy suggestions are suggested, as follows:

- The government should provide adequate rewards and incentives for promoting sustainability initiatives.
- Strict government regulations for reducing environmental footprint can boost the sustainability alliance between suppliers and manufacturers, which is beneficial for enhancing their sustainability index.
- NGOs can aid in forcing the manufacturers to apply innovative ideas for improvement in social and ecological performance.

3.8 CONCLUSIONS AND RECOMMENDATIONS

Manufacturing industries are specifically in the limelight due to the rise in pollution levels. Moreover, growing competition in the ASP industry is another prominent reason for these industries to adopt new methods for the inclusion of sustainability in their manufacturing as well as procurement activities. This has led several ASP manufacturing enterprises to explore the possibility of supply chain collaboration with their suppliers to effectively manage raw materials and components.

In this direction, the primary requirement for any ASP manufacturing company is to understand the list of MFs that put pressure on implementing sustainable collaboration with suppliers. A total of 9 MFs were selected for this study based on interactions with various industry and academic experts. The present study used the recently developed methodology of BWM for prioritizing the vital MFs for the inclusion of sustainability based on the opinion of the expert panel. The case of the manufacturer of 4-wheeler spare parts situated in northern India is considered for validating the study. The best MF was found to be "Continued pressure from NGOs to reduce waste" (MF4) while the worst was "Information and resource sharing to attract customers" (MF2) based on the consensus of the expert panel. Hence, this study provides a framework for the assessment of MFs for a sustainable collaborative alliance with suppliers in the ASP industry.

REFERENCES

Amindoust, A., Ahmed, S., Saghafinia, A., & Bahreininejad, A. (2012). Sustainable supplier selection: A ranking model based on fuzzy inference system. *Applied Soft Computing*, 12(6), 1668–1677.

Cao, M., Vonderembse, M. A., Zhang, Q., & Ragu-Nathan, T. S. (2010). Supply chain collaboration: Conceptualisation and instrument development. *International Journal of Production Research*, 48(22), 6613–6635.

Cao, M., & Zhang, Q. (2011). Supply chain collaboration: Impact on collaborative advantage and firm performance. *Journal of Operations Management*, 29(3), 163–180.

Chen, Y. J., Wu, Y. J., & Wu, T. (2015). Moderating effect of environmental supply chain collaboration: Evidence from Taiwan. *International Journal of Physical Distribution & Logistics Management*, 45(9/10), 959–978.

Gazzola, P., Del Campo, A. G., & Onyango, V. (2019). Going green vs going smart for sustainable development: *Quo vadis? Journal of Cleaner Production*, 214, 881–892.

Govindan, K., Darbari, J. D., Agarwal, V., & Jha, P. C. (2017). Fuzzy multi-objective approach for optimal selection of suppliers and transportation decisions in an eco-efficient closed loop supply chain network. *Journal of Cleaner Production*, 165, 1598–1619.

Govindan, K., Jha, P. C., Agarwal, V., & Darbari, J. D. (2019). Environmental management partner selection for reverse supply chain collaboration: A sustainable approach. *Journal of Environmental Management*, 236, 784–797.

Hsueh, C. F. (2015). A bilevel programming model for corporate social responsibility collaboration in sustainable supply chain management. *Transportation Research Part E: Logistics and Transportation Review*, 73, 84–95.

Kache, F., & Seuring, S. (2014). Linking collaboration and integration to risk and performance in supply chains via a review of literature reviews. *Supply Chain Management: An International Journal*, 19(5/6), 664–682.

Kara, M., & Firat, S. (2016). Sustainable Supplier Evaluation and Selection Criteria. *Social and Economic Perspectives on Sustainability*, Erdoğdu, M.M., Mermod, A.Y., Aşkun Yıldırım, O.B., Eds., IJOPEC Publication, London, 159–168.

Klassen, R. D., & Vachon, S. (2003). Collaboration and evaluation in the supply chain: The impact on plant-level environmental investment. *Production and Operations Management*, 12(3), 336–352.

Kushwaha, G. S., & Sharma, N. K. (2016). Green initiatives: A step towards sustainable development and firm's performance in the automobile industry. *Journal of Cleaner Production*, 121, 116–129.

Large, R. O., & Thomsen, C. G. (2011). Drivers of green supply management performance: Evidence from Germany. *Journal of Purchasing and Supply Management*, 17(3), 176–184.

Mathiyazhagan, K., Sengupta, S., & Mathivathanan, D. (2019). Challenges for implementing green concept in sustainable manufacturing: A systematic review. *OPSEARCH*, 56(1), 32–72.

Morgan, T. R., Richey Jr, R. G., & Autry, C. W. (2016). Developing a reverse logistics competency: The influence of collaboration and information technology. *International Journal of Physical Distribution & Logistics Management*, 46(3), 293–315.

Niesten, E., Jolink, A., de Sousa Jabbour, A. B. L., Chappin, M., & Lozano, R. (2017). Sustainable collaboration: The impact of governance and institutions on sustainable performance. *Journal of Cleaner Production*, 155, 1–6.

Nyaga, G. N., Whipple, J. M., & Lynch, D. F. (2010). Examining supply chain relationships: do buyer and supplier perspectives on collaborative relationships differ? *Journal of Operations Management*, 28(2), 101–114.

Ramanathan, U., & Gunasekaran, A. (2014). Supply chain collaboration: Impact of success in long-term partnerships. *International Journal of Production Economics*, 147, 252–259.

Rezaei, J. (2015). Best-worst multi-criteria decision-making method. *Omega*, 53, 49–57.

Rezaei, J. (2016). Best-worst multi-criteria decision-making method: Some properties and a linear model. *Omega*, 64, 126–130.

Sakka, O., Millet, P. A., & Botta-Genoulaz, V. (2011). An ontological approach for strategic alignment: A supply chain operations reference case study. *International Journal of Computer Integrated Manufacturing*, 24(11), 1022–1037.

Sancha, C., Gimenez, C., & Sierra, V. (2016). Achieving a socially responsible supply chain through assessment and collaboration. *Journal of Cleaner Production*, 112, 1934–1947.

Sandberg, E. (2007). Logistics collaboration in supply chains: practice vs. theory. *International Journal of Logistics Management*, 18(2), 274–293.

Simatupang, T. M., & Sridharan, R. (2002). The collaborative supply chain. *The International Journal of Logistics Management*, 13(1), 15–30.

Vachon, S., & Klassen, R. D. (2006). Extending green practices across the supply chain: The impact of upstream and downstream integration. *International Journal of Operations & Production Management*, 26(7), 795–821.

Van der Vaart, T., & van Donk, D. P. (2008). A critical review of survey-based research in supply chain integration. *International Journal of Production Economics*, 111(1), 42–55.

Zhang, M., Pawar, K. S., & Bhardwaj, S. (2017). Improving supply chain social responsibility through supplier development. *Production Planning & Control*, 28(6–8), 500–511.

CHAPTER 4

Innovative Procurement Practices Increase Productivity

Indian Dairy Industries

Yogesh Kumar Sharma, Pravin P. Patil, Neema Tufchi, and Yaşanur Kayıkcı

CONTENTS

4.1 INTRODUCTION

Since 1998, India has been one of the major manufacturers and buyers of dairy products, with continuous development in the accessibility of milk and dairy products. Dairy farming and other actions play an important role in the rural Indian economy by providing employment. India has one of the world's largest populations, yet its milk production is considerably lower in comparison with other dairy-producing countries.

In India, the majority of dairy product is consumed by its producers; the rest is sold in the form of liquid milk. The dairy industry plays a crucial role in the growth of the Indian economy (Mangla et al., 2019b; Singh et al., 2019).

Many people find employment through dairy industries, thus helping in the reduction of poverty in the rural areas of India. The Indian dairy industry has the potential to enhance development of the country (Yawar and Kauppi, 2018; Ishrat et al., 2018). India is one of the world's most populous countries, with 1.30 billion people, and most of the population depend on dairy products like milk, curd, cheese, sweets, etc. To meet demand, dairy products industry is fully reliant on the raw material i.e., milk, which is a perishable with wide cyclic fluctuations (Ciarli et al., 2018; Ali et al., 2019). Now, the questions that arise are how and from where to collect milk to meet the daily needs of the population, or what factors influence milk procurement (Lehtinen, 2012). India is one of the largest milk producers in the world but there is still a huge gap between supply and demand. If this gap is narrowed by improving the milk supply chain, it would directly enhance the economy of our country (Kamath et al., 2019). In improving the supply of milk, sustainability is very important throughout the whole supply chain. To deal with the universal challenges, the concept of "sustainability" has an impact on the milk supply and dairy sector, as emphasises the consumption of resources in the most effective way (WCED, 1987). According to the Indian government's economic survey 2015–2016, India claims nearly 146.3 million tonnes of yearly milk production.

In the current study, attempts have been made to investigate innovative procurement practices in increasing the productivity of Indian dairy industries (Rijswijk and Brazendale, 2017; Glover and Poole, 2019).

Milk procurement is achieved using the latest technologies and innovations in dairy industry, such as in the following areas:

Cattle health—Milk production from animals is directly dependent on their wellbeing (Burkitbayeva et al., 2019). Thus, it is very important to track the health of cows or buffalos for good production of milk. This can be done with the use of smartwatches that are easily available in the market. These watches continuously monitor pulse rate and other signs in the animals for the examination of their health.

Irregularities in milk production—Bluetooth Low Energy (BLE) and radio frequency identification (RFID) technologies are used to track cows. All animals are given a unique identification number to track the animal's health and milk production (Gupta, 2017; Gargiulo et al., 2018; Sharma et al., 2018c). Milk production depends on the food which animals eat, affecting the taste and quality of milk. Feed is an important factor in milk production and constitutes a major cost in the dairy's operation. Thus, the quality of feed should be good to ensure greater yield both in quantity and quality of milk (Hall et al., 2019).

Based on a newly published report by IMARC Group titled "Dairy Industry in India 2019 Edition: Market Size, Growth, Prices, Segments, Cooperatives, Private Dairies, Procurement and

Distribution" the dairy marketplace in India had grown to a value of INR 9.168 billion in 2018.

4.2 MOTIVATION AND AIMS

The relatively low per-capita availability of milk and the poorly remunerative dairy procurement methods in India are the main objectives of this study. In India, most milk is wasted due to lack of awareness, poor technologies and policies, and inefficient cold chain (Sharma et al., 2018a, 2018b). In the current chapter, more stress is thus placed on the procurement of milk by innovative practices in the dairy supply chain aimed at increasing its productivity. This will help in establishing or making new policies and plans for successful execution and improvement of the dairy sector in India.

4.3 LITERATURE REVIEW

In this literature review, focus is mainly on the three key factors, i.e., dairy supply chain and food, functional problems affecting milk production, and input cost, i.e., grounded on milk procurement price. Approximately 80% of Indian milk is produced by villagers and is handled by the unstructured sector and the remaining 20% is handled by the structured sector (Rajendran and Mohanty, 2004). Eastwood et al. (2016), Rijswijk and Brazendale (2017), and Eastwood et al. (2018) highlight the numerous issues faced by dairy farmers. The two-axis gaining pricing process does not reflect the production cost of milk when payment is credited to a farmer's account (Saravanakumar and Jain, 2009). Mor et al. (2019b) mainly focused on the critical factors of the dairy supply chain. Prasad and Kumari (2016) focused attention on rethinking cooperatives for sustainable development. Meganathanet et al. (2010) suggested that the main concern for milk producers are profitless prices for cattle goods. Sharma et al. (2018a, 2018b) and Mangla et al. (2019a) mostly focused on the implementation of sustainability in food supply chains. Läpple and Thorne (2019) investigated how innovation impacts economic sustainability and is an important step in attaining more sustainable production development in the dairy sector. Soteriades et al. (2016)

found that the environmental "sustainability" of concentrated dairy farming depends on farming systems and situations. Kulandaiswamy (1982) recommended a cost-based value-obtaining system to fix the cost of input. Thus, to fix the cost of input, which is rather unstable, is an important step toward sustainability. According to Subburaj et al. (2015) and D'Haene et al. (2019), milk procurement prices should be fixed on the basis of the cost of production of milk, cyclic variations, and overall market movements.

Based on the literature review, current research has mainly focused on innovative procurement practices and policies in Indian dairy industries that are aimed at increasing productivity and enhancing the economy of India (Sharma, 2015; Eastwood et al., 2016; Asayehegn et al., 2019).

4.4 INNOVATIVE PRACTICES IN THE INDIAN DAIRY INDUSTRY

India is the largest manufacturer and largest buyer of milk or dairy products in the world. The middle class population in India has increased rapidly, so it will face a shortage of milk or milk products (Mangla et al., 2016; Kumar and Mohan, 2018; Sharma et al., 2019). The NDP-I (National Dairy Plan Phase I) is a central sector scheme of the government of India and is supported by the National Dairy Support Project (NDSP). The objective of the organization is to increase milk productivity and market access for farmers so they can fulfill growing demands. The National Dairy Development Board (NDDB) is the chief executing agency for the NDP-I. These organizations are mainly focused on the improvement of the following milk procurement practices in the Indian dairy market.

4.4.1 Supply Chain Consolidation for Village-Based Milk Procurement

Presently in India, there are almost 16 million milk producers who belong to approximately 150,000 village dairy cooperatives. As milk is a highly perishable item, any errors in the supply chain may spoil the raw milk before it can reach processing plants. It is very important to maintain proper hygiene while bringing milk to collection centers (Subburaj et al., 2015).

To address the above challenges, the NDSP has financed environmental impact assessments (EIAs) and the purchase and installation of bulk milk chillers (BMCs) at village milk collection points, which generate savings in transportation, operations, handling, and processing costs. To streamline milk collection and test for the quality of the supplied milk, EIAs also purchased standardized automated milk collection units (AMCU) and data processor-based milk collection units (DPMCU) for collection centers, along with associated IT systems. Adulteration testing kits were also supplied. These EIAs also procured milk cans and provided them to producers to ensure proper hygiene during the transport of milk to collection centers. The NDSP also encouraged the formation of new village-level cooperatives to bring in new producers who could supply milk. Since the mid-point of the project's implementation period, 23,487 villages have been covered and 660,935 additional milk producers have been introduced.

4.4.2 Framework Agreements for Decentralized Procurement of Equipment

In India, there are about 150 EIAs working in 18 different states that frequently require some standard items (for example, BMCs). To overcome the issues of purchasing, delays in order placement, and delays in the release of payments, the Indian government has proposed Framework Agreements (FAs) for such items. The use of FAs has not only resulted in the acceleration of the procurement process but also in monetary savings of up to 15% in many cases.

4.4.3 Procurement Error and Quality Assurance

The government also engaged the services of a trustworthy check-up agency for quality control of items procured by FA arrangements. A quality assurance plan (QAP) was developed for each item to minimize the scope of absences in quality, and inspections start right from the manufacturing stage of equipment.

Besides the rectification and detection of defects in items by the supplier, these proactive measures have resulted in the cancellation of FAs with suppliers who were unable to maintain required quality standards. A web-based procurement management information system (MIS) was developed to help in monitoring and utilization of the overall progress on decentralized procurement. The MIS also helped EIAs in sharing data among each other.

4.5 ENDORSEMENTS

There are many suggestions for policymakers and higher authorities based on procurement efforts currently being carried out in the Indian dairy sector, as follows:

- The formation of special dairy zones
- The implementation of self-motivated milk procurement methods
- The establishment of cooperative societies
- The construction of feed banks and the increasing of feed productivity

4.5.1 Formation of Special Dairy Zones

As we all know, the demand for milk increases day by day, so to fulfill demand and improve the production of milk and dairy products, a special dairy zone (SDZ) should be formed. The SDZ method is very significant in terms of increasing the production and distribution of milk. If surplus milk is produced, it can be exported, thereby contributing to foreign trade.

4.5.2 Implementation of Self-Motivated Milk Procurement Methods

The two-axis theory explains that the cost of milk is formulated by fixing a pre-determined amount for fat and solids-not-fat (SNF). In the current system, fat and SNF are usually given equal prices and the per kg prices for fat and SNF will be fixed. This method is normally used for milk procurement, but it is not capable of giving proper payment to milk manufacturers in India since the production cost is not considered in this method.

The milk procurement price might be reviewed 10% each year because a dairy agronomist devotes nearly 60% of the entire working cost to fodder alone. The fodder price spontaneously increases by 5% half-yearly. The complete milk production rate also rises by 10–10.5% percent each year. This will allow steady financial profit to dairy agronomists.

4.5.3 Establishment of Cooperative Societies

A milk co-operative society should make clear rules since it provides the link between producers and buyers of milk. Cooperative societies are not profit groups. The chief goal of cooperative society is to fulfill the desires of both the makers and buyers. If the members of cooperative societies are chosen by dairy agronomists, the probabilities of loss are decreased (Prasad and Kumari, 2016)

4.5.4 Construction of Feed Banks and Increasing of Feed Productivity

The key purpose of feed banks are "feed for all cattle." A cattle feed card may be discreetly distributed only to the dairy cooperative society's driving associates.

The remains of food grains, and feed should be distributed to members of dairy cooperatives at subsidized rates. Thus, the cost of milk production can be reduced for farmers who are dependent on the market to procure feed. In this study it was observed that farmers want additional official support for growing feed and delivering it to milk producers through dairy cooperatives. Using this technique, the cost price of feed can be decreased.

Consideration should be given to fodder production and farmers should be appreciated for the cultivation of fodder through cluster farming. Water and farming land should also be protected from exploitation. Combined projects between cooperative societies and village panchayat in feed farming may provide support to powerless dairy agriculturalists.

4.6 CONCLUSION

The current chapter is projected to narrow the gap regarding the procurement of milk in the Indian dairy industry by implementing innovative practices. In addition to this, the present chapter provides recommendations for decision-makers for the successful procurement of milk. It is noted that 75% of perishable food is wasted due to the negligence of sellers. The outcome of the chapter is that the procurement of milk in daily supply can be increased by the adoption of innovative practices. Procurement practices are highly affected by irregular wastage and poor handling of milk at the processing site as well as at the source (Mor et al., 2018b, 2019a). More wastage can happen due to multiple collection points and unhygienic practices. These hurdles further increase the pressure on the Indian dairy industry for substantial development in their procurement systems (Mor et al., 2018a, 2019b; Mangla et al., 2019a). However, the adoption of newer technologies will help improve productivity. According to the dairy minister in India, the Indian dairy sector has increased at a rate of 6.4% annually over the last four years against a global growth rate of 1.7 %, and the aim currently is to improve milk productivity per animal. The minister also drew attention to the adoption of new technologies under the Rashtriya Gokul Mission, such as embryo transfer technology, creation of facility for sex-sorted semen production, and genomics selection, which would help improve the productivity. Additionally, the tackling of milk adulteration is a key factor in the procurement of milk at village level. In this regard, the National Programme for Dairy Development (NPDD) is an important agency in maintaining trust among farmers as well as customers. In addition, many studies have found that in the dairy industry, a lack of refrigeration and defective power supply are major sources of food wastage, which require the implementation of innovative refrigeration technologies to reduce spoilage and improve milk quality. Apart from achieving the production and consumption of safe milk, established laboratories in dairy plants would help to encourage export. India accounts for only 0.01% of the worldwide dairy export market. The dairy sector is trying to build long-term sustainable growth. Existing financial models calculate the increased costs for the dairy sector will be handled by escalating overseas market access in combination with value addition through technical revolution across the supply chain.

Organizations must change their traditional dairy practices in the supply chain. To attain sustainability in the dairy supply chain, it is compulsory to minimize waste at all levels of supply chain, including the packaging, transportation, processing, storage, and distribution of processed milk products.

For a sustainable dairy industry, the focus should be mainly on new farming practices that improve sustainability. Research has been done refining farm facilities so that farmers may handle milk to meet a transformed dairy supply chain (Augustin et al., 2013). Technological innovations, automatic milk testing, effective collection centers, traceability in transportation and quality issues, and the effective implementation of information technology system can help the dairy sector to achieve their long-term goals. Therefore, the proposed innovative practices in daily industry will help dairy experts to plan their procurement practices competently to fulfill the product quality and sustainability standards.

4.7 FUTURE SCOPE

The findings of the present work depend on the support of academics and policymakers from the dairy industry. The projected innovative techniques can be useful in other milk procuring and processing firms both across India and in other countries. The proposed innovative techniques may also be implemented in other perishable food processing industries.

ACKNOWLEDGMENT

The authors acknowledge and express the gratitude for the support of the research facilities and funds provided by the Department of Mechanical Engineering, Graphic Era (Deemed to be) University, Dehradun, India.

REFERENCES

Ali, S. M., Moktadir, M. A., Kabir, G., Chakma, J., Rumi, M. J. U., & Islam, M. T. (2019). Framework for evaluating risks in food supply chain: Implications in food wastage reduction. *Journal of Cleaner Production*, *228*, 786–800.

Asayehegn, K., Temple, L., Vaast, P., & Iglesias, A. (2019). Innovation systems to adapt to climate change: Lessons from the Kenyan coffee and dairy sectors (pp. 1–24). In W. L. Filho, ed., *Handbook of Climate Change Resilience*. Springer Nature, Cham, Switzerland.

Augustin, M. A., Udabage, P., Juliano, P., & Clarke, P. T. (2013). Towards a more sustainable dairy industry: Integration across the farm–factory interface and the dairy factory of the future. *International Dairy Journal*, *31*(1), 2–11.

Burkitbayeva, S., Janssen, E., & Swinnen, J. (2019). Technology adoption and value chains in developing countries: Panel evidence from dairy in Punjab. *LICOS Discussion Paper Series*, 1–51.

Ciarli, T., Savona, M., Thorpe, J., & Ayele, S. (2018). Innovation for inclusive structural change. A framework and research agenda. *A Framework and Research Agenda (January 23, 2018). SWPS, 4.*

D'Haene, E., Desiere, S., D'Haese, M., Verbeke, W., & Schoors, K. (2019). Religion, food choices, and demand seasonality: Evidence from the Ethiopian milk market. *Foods*, *8*(5), 167.

Eastwood, C. R., Greer, J., Schmidt, D., Muir, J., & Sargeant, K. (2018). Identifying current challenges and research priorities to guide the design of more attractive dairy-farm workplaces in New Zealand. *Animal Production Science*, [online], doi: 10.1071/AN18568.

Eastwood, C. R., Jago, J. G., Edwards, J. P., & Burke, J. K. (2016). Getting the most out of advanced farm management technologies: Roles of technology suppliers and dairy industry organisations in supporting precision dairy farmers. *Animal Production Science*, *56*(10), 1752–1760.

Gargiulo, J. I., Eastwood, C. R., Garcia, S. C., & Lyons, N. A. (2018). Dairy farmers with larger herd sizes adopt more precision dairy technologies. *Journal of Dairy Science*, *101*(6), 5466–5473.

Glover, D., & Poole, N. (2019). Principles of innovation to build nutrition-sensitive food systems in South Asia. *Food Policy, 82*, 63–73.

Gupta, S. (2017). Dairy industries in India: Technological implementation and challenges. *Journal of HR, Organizational Behaviour & Entrepreneurship Development*, *1*(2), 1–7.

Hall, A., Turner, L., & Kilpatrick, S. (2019). Using the theory of planned behaviour framework to understand Tasmanian dairy farmer engagement with extension activities to inform future delivery. *The Journal of Agricultural Education and Extension*, *25*(3), 195–210.

Ishrat, S. I., Grigg, N. P., Jayamaha, N., & Pulakanam, V. (2018). Cashmere industry: Value chains and sustainability. In C. K. Y. Lo & J. Ha-Brookshire, eds., *Sustainability in Luxury Fashion Business* (pp. 113–132). Springer, Singapore.

Kamath, V., Biju, S., & Kamath, G. (2019). A Participatory Systems Mapping (PSM) based approach towards analysis of business sustainability of rural Indian milk dairies. *Cogent Economics & Finance*, *7*(1), 1–16.

Kulandaiawamy, V. (1982). *Cooperative Dairying in India*, Rainbow Publication, First Edition, Chapter IV, pp. 125–145.

Kumar, D., & Mohan, A. (2018). Factors leading to customer satisfaction in dairy industry: A study in Indian perspective. *International Journal on Customer Relations*, 6(1), 21–30.

Läpple, D., & Thorne, F. (2019). The role of innovation in farm economic sustainability: Generalised propensity score evidence from Irish dairy farms. *Journal of Agricultural Economics*, *70*(1), 178–197.

Lehtinen, U. (2012). Sustainability and local food procurement: A case study of Finnish public catering. *British Food Journal*, *114*(8), 1053–1071.

Mangla, S. K., Luthra, S., Jakhar, S. K., Kumar, A., & Rana, N. P. (2019a). *Sustainable Procurement in Supply Chain Operations*. CRC Press.

Mangla, S. K., Sharma, Y. K., Patil, P. P., Yadav, G., & Xu, J. (2019b). Logistics and distribution challenges to managing operations for corporate sustainability: Study on leading Indian diary organizations. *Journal of Cleaner Production*, *238*, 117620.

Mangla, S. K., Sharma, Y. K., & Patil, P. P. (2016). Using AHP to rank the critical success factors in food supply chain management. *Int. Conf. on Smart Strategies for Digital World-Industrial Engineering Perspective*, Nagpur, India. 58.

Meganathan, N., Selvakumar, K. N., Prabu, M., Pandian, A. S. S., & Kumar, G. S. (2010). Constraint analysis of tribal livestock farming in Tamil Nadu. *Tamilnadu Journal of Veterinary and Animal Sciences*, 6(1), 12–18.

Mor, R. S., Bhardwaj, A., & Singh, S. (2018a). Benchmarking the interactions among performance indicators in dairy supply chain: An ISM approach. *Benchmarking: An International Journal, 25*(9), 3858–3881.

Mor, R. S., Singh, S., & Bhardwaj, A. (2018b). Exploring the causes of low-productivity in dairy supply chain using AHP. *Jurnal Teknik Industri, 19*(2), 83–92.

Mor, R. S., Bhardwaj, A., Singh, S., & Kharub, M. (2019a). Framework for measuring the procurement performance in the dairy supply chain. In S. K. Mangla, S. Luthra, S. K. Jakhar, A. Kumar, N. P. Rana, eds., *Sustainable Procurement in Supply Chain Operations*. CRC Press, Boca Raton, FL, 61.

Mor, R. S., Bhardwaj, A., Singh, S., & Nema, P. K. (2019b). Framework for measuring the performance of production operations in the dairy industry. In J.C. Essila, ed., *Managing Operations Throughout Global Supply Chains* (pp. 20–49). IGI Global, Pennsylvania, PA.

Prasad, C. S., & Kumari, J. (2016). Rethinking cooperatives for sustainable development: Insights from Vasudhara Dairy and Dharani Organic cooperatives. In *11th ICAAP Cooperative Research Conference, New Delhi. "Cooperatives and Sustainable Development"*. http://www.ica-ap.coop/sites/ica-ap.coop/files/Shambu_Prasad India.PDF.

Rajendran, K., & Mohanty, S. (2004). Dairy co-operatives and milk marketing in India: Constraints and opportunities. *Journal of Food Distribution Research, 35*(856-2016-56967), 34–41.

Rijswijk, K., & Brazendale, R. (2017). Innovation networks to stimulate public and private sector collaboration for advisory services innovation and coordination: The case of pasture performance issues in the New Zealand dairy industry. *The Journal of Agricultural Education and Extension, 23*(3), 245–263.

Saravanakumar, V., & Jain, D. K. (2009). Evolving milk pricing model for agribusiness centres: An econometric approach. *Agricultural Economics Research Review, 22*(347-2016-16735), 155–160.

Sharma, V. P. (2015). Determinants of small milk producers' participation in organized dairy value chains: Evidence from India. *Agricultural Economics Research Review, 28*(347-2016-17180), 247–261.

Sharma, Y. K., Mangla, S. K., Patil, P. P., & Liu, S. (2019). When challenges impede the process: For circular economy-driven sustainability practices in food supply chain. *Management Decision, 57*(4), 995–1017.

Sharma, Y. K., Mangla, S. K., Patil, P. P., & Uniyal, S. (2018a). Analyzing sustainable food supply chain management challenges in India. In R. Mangey, & D. J. Paulo, eds., *Soft Computing Techniques and Applications in Mechanical Engineering* (pp. 162–180). IGI Global, Pennsylvania, PA.

Sharma, Y. K., Mangla, S. K., Patil, P. P., & Uniyal, S. (2018b). Sustainable food supply chain management implementation using DEMATEL approach. In N. A. Siddiquis, S. M. Tauseef, & K. Bansal, eds., *Advances in Health and Environment Safety* (pp. 115–125). Springer, Singapore.

Sharma, Y. K., Mangla, S. K., Patil, P. P., Yadav, A. K., Jakhar, S. K., & Luthra, S. (2018c). Ranking the IT based technologies to enhance the safety and security of food using AHP approach. *Mathematics Applied in Information Systems, 2*, 108–122.

Singh, R. K., Luthra, S., Mangla, S. K., & Uniyal, S. (2019). Applications of information and communication technology for sustainable growth of SMEs in India food industry. *Resources, Conservation and Recycling, 147*, 10–18.

Soteriades, A. D., Stott, A. W., Moreau, S., Charroin, T., Blanchard, M., Liu, J., & Faverdin, P. (2016). The relationship of dairy farm eco-efficiency with intensification and self-sufficiency. Evidence from the French dairy sector using life cycle analysis, data envelopment analysis and partial least squares structural equation modelling. *PloS one, 11*(11), e0166445.

Subburaj, M., Babu, T. R., & Subramonian, B. S. (2015). A study on strengthening the operational efficiency of dairy supply Chain in Tamilnadu, India. *Procedia-Social and Behavioral Sciences, 189*, 285–291.

WCED. (1987). *Our Common Future: Report of the World Commission on Environment and Development.* Oxford University Press, Oxford, UK.

Yawar, S. A., & Kauppi, K. (2018). Understanding the adoption of socially responsible supplier development practices using institutional theory: Dairy supply chains in India. *Journal of Purchasing and Supply Management, 24*(2), 164–176.

Index

Printed in the United States
by Baker & Taylor Publisher Services